T0077239

A Short Introduction to Mathematical Concepts in Physics

Mathematics is the language of physics and yet, mathematics is an enormous subject. This textbook provides an accessible and concise introduction to mathematical physics for undergraduate students taking a one semester course.

It assumes the reader has studied a year of introductory physics and three semesters of basic calculus, including some vector calculus, but no formal training in differential equations or matrix algebra. It equips readers with the skills and foundational knowledge they need for courses that follow in classical mechanics, electromagnetism, quantum mechanics, and thermal physics.

This book exposes students early on to the kinds of mathematical manipulations they will need in upper-level courses in physics. It can also serve as a useful reference for their further studies.

Key features:

- Accompanied by homework problems and a solutions manual for instructors, available upon qualifying course adoption
- Bridges the gap between calculus and physics, explaining fundamental mathematics (differentiation, integration, infinite series) in physical terms
- Explores quick extensions into mathematics useful in physics, not typically taught in math courses, including the Gamma Function, hyperbolic functions, Gaussian integrals, Legendre polynomials, functions of a complex variable, and probability distribution functions

A Short Introduction to Mathematical Concepts in Physics

Jim Napolitano

CRC Press is an imprint of the
Taylor & Francis Group, an **informa** business

Designed cover image: © Jim Napolitano

First edition published 2024
by CRC Press
2385 NW Executive Center Drive, Suite 320, Boca Raton FL 33431

and by CRC Press
4 Park Square, Milton Park, Abingdon, Oxon, OX14 4RN

CRC Press is an imprint of Taylor & Francis Group, LLC

ISBN: 978-1-032-40977-1 (hbk)
ISBN: 978-1-032-40430-1 (pbk)
ISBN: 978-1-003-35565-6 (ebk)

DOI: 10.1201/9781003355656

Typeset in Nimbus Roman
by KnowledgeWorks Global Ltd.

Publisher's note: This book has been prepared from camera-ready copy provided by the authors.

Dedication

For my students in PHYS 2502 in Spring 2022 and Spring 2023, whose hard work and enthusiasm made this book a fun project

Contents

Preface

I happily agreed to teach our sophomore "math methods" course in the Spring 2022 semester, but it was somewhat daunting. Not only had I never taught this course before, I'd never even taken one, either as an undergraduate or graduate student. In the 1970s, it seems, we had a lot more required mathematics courses as physics majors, and we ended up filling in what more we needed by ourselves when we needed the necessary math for our physics courses.

My goal was to teach a course that introduced students not only to most of the mathematics they would see in their future physics courses but also to attach a physical context to the three calculus courses (through multivariable calculus) they had taken as prerequisites. There are many excellent textbooks on mathematical methods in physics, but I was worried that I would have to pick and choose from different chapters of any of them in order to meet the flow of material I wanted to present. This made me concerned that I would miss including material that other authors felt was prerequisite, and I didn't want to confuse and frustrate my students.

Consequently, I decided to bite the bullet and write up my own notes for the course, each week for the material that would be covered the following week. I was pleased that students responded well to these notes, finding them useful for following and reinforcing what we covered in class. After many embellishments, I used the notes for the same course in Spring 2023 and used that experience to fix some mistakes and smooth out plenty of rough edges. This textbook is the final result.

I often reminded students that ours was a physics course, not a math course, and this point of view is reflected in the book. Almost no proofs are rigorous in any sense, and I often rely on readers' physical insight to approach solutions. That said, I also often encouraged students to take more formal math courses in topics like differential equations, linear algebra, and complex variables, where they will see the importance of rigor and also get drilled in the many techniques which we only had time to lightly cover in class.

My class covered pretty much the entire book in one semester. I resisted the urge to put in more material as I prepared the manuscript, but some slipped in, mostly as examples of physics problems. Some instructors might opt to skip all or parts of the later chapters in favor of covering these and other examples. Also, while teaching the course, I came to realize that some fundamental concepts that I assumed students had seen before, such as complex numbers and Taylor series expansions, were not uniformly covered in their previous courses. As a result, some instructors may want to cover these important and foundational concepts in more detail.

My teaching style is to write things on the board in real time, and this led me to use some potentially unconventional notations in the book. For example, I use arrows over variable names to represent spatial vectors, single underlines to represent generalized column vectors, and double underlines for matrices, all because these are easy to write on the board, and I wanted students to be able to follow along in the book without having to figure out how to translate the notation.

At Temple University, this course is accompanied by a laboratory where students solve problems with computer applications. This is reflected in some of the end-of-chapter exercises, and in some of the essential text, although I do not necessarily call out the need for computers in a particular exercise. I believe that students should develop their own sense of when it is appropriate, or necessary, to use a computer to solve a problem, including the use of symbolic manipulation applications. I use MATHEMATICA almost exclusively for these things, but some students are more proficient in MATLAB, MAPLE, PYTHON or C++ or other applications or languages, and that is of course up to them.

As a disclaimer, I want to point out that I have a MATHEMATICA textbook also published by CRC/Taylor & Francis. My course notes mentioned this application in many places. I tried to scrub this textbook of any specific references to MATHEMATICA, but I may have missed some.

I want to give many thanks to the wonderful editors and technical staff at CRC/Taylor & Francis. A special thanks to Rebecca Hodges-Davies, Commissioning Editor for Physics, who originally contacted me about projects I might have in mind and who ran with the idea of a book based on my class notes. I also want to thank Danny Kielty, Senior Editorial Assistant in Physics, who shepherded me through the completion of the manuscript.

Jim Napolitano
June 2023
Philadelphia

Author Biography

Jim Napolitano is Professor of Physics at Temple University. His undergraduate degree is in Physics from Rensselaer Polytechnic Institute, and he earned his PhD in Physics at Stanford University. For the first ten years of his career, he was on the staffs at Argonne National Laboratory and Jefferson Laboratory, but has been a Physics faculty member at Rensselaer and Temple for more than 25 years.

Professor Napolitano has taught courses at all level, from introductory physics, to intermediate and upper level theoretical and experimental physics, and graduate quantum mechanics. His research field is Experimental Nuclear Physics, and has published many papers in the Physical Review, Physical Review Letters, and other journals. In 2016 he shared the Fundamental Physics Breakthrough Prize with his collaborators on the Daya Bay Reactor Neutrino Experiment. He was elected a Fellow of the American Physical Society in the Division of Nuclear Physics in 2011, has contributed as a member of the Physical Review C Editorial Board, and serves on or chairs several review panels in the field of Nuclear and High Energy Physics.

1 Basic Concepts

Mathematics is the language of physics. According to lore, Newton invented calculus in order to explain his philosophies to weaker minds. Oliver Heaviside invented vector calculus to cast Maxwell's Equations into a form that made calculations so much easier. Eugene Wigner introduced group theory into Quantum Mechanics so that physicists had a framework for exploiting symmetry in nature.

Consequently, it is of the highest importance for students of physics to be well versed in many different branches of mathematics. This course should not be considered to cover "all" the mathematics you will need, but hopefully it will introduce you to the most important concepts that you'll see later.

This first chapter is meant to cover the very basic ideas, which will lay the groundwork for the rest of the course. Most everything in this chapter is a review. But not everything.

1.1 FUNDAMENTALS

One very important and simple thing to hit home is the idea of variable names as "dummies." That is, I will use things like x, y, z, u, v, w,... a lot, but they have no physical meaning until I tell you they do. And their physical meaning will mostly be different for different problems. Also, there is often no standard terminology, so one physics class might use a variable to mean some thing, and another class might use a different variable to mean the same thing.

So, be careful. If you see $x(t)$, for example, then it *probably* means position as a function of time, but not necessarily. Always confirm at the outset what is the nomenclature when you're working on a physical problem.

1.1.1 NUMBERS

Numbers are used to "measure" quantities in physics. Mathematicians talk about "number fields" including integers $\mathbb{Z}$, rational numbers $\mathbb{Q}$, real numbers $\mathbb{R}$, and complex numbers $\mathbb{C}$. I will use these symbols from time to time. If I write something like $x \in \mathbb{R}$, then I mean that the variable x represents some real number.

We have some special sets of numbers. (I'll get to a more formal definition of a "set" shortly.) For example, $\mathbb{R}^2$ is the set of pairs of real numbers. Physically, you can think of $\mathbb{R}^2$ as the set of points in a plane. Similarly, $\mathbb{R}^3$ is the set of real numbers in three-dimensional space.

Addition, subtraction, multiplication, and division of integers, rational and real numbers are all exactly what you think they are. It is likely that you've already learned some things about the complex number system, but I'll give you some details next.

DOI: 10.1201/9781003355656-1

1.1.1.1 Complex numbers

The "imaginary" number i is defined as the square root of -1. That is

$$i = \sqrt{-1}$$

This makes it possible to define a "complex number" $z \in \mathbb{C}$ as

$$z = x + iy \qquad \text{where} \qquad x \in \mathbb{R}, y \in \mathbb{R}$$

(Note that $\mathbb{C}$ sounds a little like $\mathbb{R}^2$, in that there is a one-to-one correspondence through the real numbers.) We refer to x as the "real part" of z and write $x = \Re(z)$. Similarly, y is the "imaginary part" of z and write $y = \Im(z)$. On the blackboard, I will typically write $x = \mathrm{Re}\, z$ and $y = \mathrm{Im}\, z$, but these two notations mean the same thing.

The "complex conjugate" z^* of a complex number $z = x + iy$ is

$$z^* = x - iy$$

The "modulus" of a complex number $z = x + iy$ is

$$|z| = \sqrt{(x^2 + y^2)} = (x^2 + y^2)^{1/2}$$

Addition and subtraction of complex numbers just mean the addition and subtraction separately of their real and imaginary parts. Multiplication of complex numbers requires a little more care. For $z_1 = x_1 + iy_1$ and $z_2 = x_2 + iy_2$, then

$$z_1 z_2 = (x_1 + iy_1)(x_2 + iy_2) = (x_1 x_2 - y_1 y_2) + i(x_1 y_2 + x_2 y_1) \tag{1.1}$$

It is a nontrivial observation that $z_1 z_2 = z_2 z_1$, namely that complex multiplication commutes. Note also that

$$z^* z = z z^* = |z|^2$$

It is also nontrivial that the complex conjugate of a product is the product of the complex conjugates, that is

$$\begin{aligned}
(z_1 z_2)^* &= [(x_1 + iy_1)(x_2 + iy_2)]^* = [(x_1 x_2 - y_1 y_2) + i(x_1 y_2 + x_2 y_1)]^* \\
&= (x_1 x_2 - y_1 y_2) - i(x_1 y_2 + x_2 y_1) = (x_1 - iy_1)(x_2 - iy_2) = z_1^* z_2^*
\end{aligned}$$

Another nontrivial observation is that $|z_1 z_2| = |z_1||z_2|$. To show this, use (1.1) to write

$$\begin{aligned}
|z_1 z_2| &= \left[(x_1 x_2 - y_1 y_2)^2 + (x_1 y_2 + x_2 y_1)^2\right]^{1/2} \\
&= \left[x_1^2 x_2^2 + y_1^2 y_2^2 + x_1^2 y_2^2 + x_2^2 y_1^2\right]^{1/2} \\
&= \left[x_1^2 (x_2^2 + y_2^2) + y_1^2 (y_2^2 + x_2^2)\right]^{1/2} = \left[(x_1^2 + y_1^2)(x_2^2 + y_2^2)\right]^{1/2} = |z_1||z_2|
\end{aligned}$$

Division of complex numbers is best understood using the modulus, that is

$$\frac{z_1}{z_2} = \frac{z_1}{z_2}\frac{z_2^*}{z_2^*} = \frac{1}{|z_2|^2} z_1 z_2^*$$

and division of complex numbers reduces to multiplication.

For more advanced manipulations of complex numbers, for example $\sqrt{z}$, it is best to wait until Section 2.4.2 when we see how to write complex numbers in terms of the modulus and a "phase" using Euler's relation. A glimpse of how complex numbers are tied to group theory is discussed in Section 6.3.8.

1.1.2 FUNCTIONS

A "function" maps from one number system onto itself or another number system. Mostly we will deal with functions that map real numbers onto real numbers. For example

$$f : \mathbb{R} \mapsto \mathbb{R} \qquad \text{or} \qquad u = f(x)$$

where a function of one real number gives another real number, or

$$g : \mathbb{R}^2 \mapsto \mathbb{R} \qquad \text{or} \qquad u = g(x,y)$$

where a function of two real numbers gives one real number.

There will be many other examples, particularly when we get into vector functions. One very familiar example is the electric field from a charge distribution, which maps $\mathbb{R}^3$, that is, position in three-dimensional space, onto $\mathbb{R}^3$, that is, the electric field itself. In other words

$$\vec{f} : \mathbb{R}^3 \mapsto \mathbb{R}^3 \qquad \text{or} \qquad \vec{E} = \vec{f}(\vec{r})$$

Physicists use a wide variety of different notations, so don't get hung up on that. For example, if I write something like $f = f(x)$, all I mean is to emphasize that f is a function of the single variable x. Your intuition should be good enough to get you through anything confusing, but always ask questions if you're unsure.

1.1.3 SETS

A set is some collection of objects. These objects can be numbers, pictures, functions, in fact anything. We denote as symbols to represent the objects, enclosed in curly brackets, that is $\{a,b,c,\ldots\}$. Simple examples of sets are $\mathbb{R}$ and $\mathbb{Z}$.

A subset is a set contained within a larger set. We write $A \subset B$ if set A is a subset of set B. For example, $\mathbb{Z} \subset \mathbb{R}$.

If we write $A = \{a_i\}$ and $B = \{b_j\}$, where i and j enumerate the elements of A and B, then we can form the set $A \times B = \{a_i, b_j\}$ which is known as a Cartesian products. A simple example is $\mathbb{R} \times \mathbb{R} = \mathbb{R}^2$.

Functions can in fact map from any set onto any other set, but we're not going to get that fancy in this course.

1.1.4 GROUPS

We can't get into it much in this course, but I want to at least introduce you to the concept of a "group." Groups are fundamental in formulating physical theory. In the same way that "numbers" measure "quantities," groups are used to measure "symmetry" in nature. One of the most common examples is the group of 3×3 real orthogonal matrices[1] with unit determinant, which physicists write as SO(3), that measures the rotational symmetry of the three-dimensional world.

In order to define a group G, you need first to have two things, namely a set $\{x, y, z, \dots\}$ and a binary operation $\circ$ between elements of the set. These form a group if the following three "group axioms" are satisfied:

1. There is an element $\mathbb{1}$ of G, called the *identity element* such that for any x in G, $x \circ \mathbb{1} = x = \mathbb{1} \circ x$.
2. For any x that is an element of G, there exists an inverse element x^{-1}, also in G, such that $x \circ x^{-1} = \mathbb{1} = x^{-1} \circ x$
3. For any elements x, y, z in G, $x \circ (y \circ z) = (x \circ y) \circ z$. That is, the binary operation is associative.

Note that there is no requirement that the binary operation be commutative, that is $x \circ y = y \circ x$. (Think about matrix multiplication, for example.) If this commutative property holds, then we say the group is "Abelian."

One simple example of a group is the real numbers $\mathbb{R}$ under addition. The identity element is zero, the inverse of any x is $-x$, and addition is clearly associative. This is also an Abelian group. The integers $\mathbb{Z}$ also form a group under addition, and in fact are a "subgroup" of $\mathbb{R}$.

A slightly more advanced group is the real numbers, excluding zero, under multiplication. Let's call that group $\mathbb{R}^*$. We have to exclude zero because if I divide any other element in the group by zero, I don't get a real number. For $x \in \mathbb{R}^*$, we see easily that $1 = \mathbb{1}$ and $x^{-1} = 1/x$. Multiplication of real numbers is obviously associative. This is also an Abelian group.

Things get more interesting when we discuss matrix operations. See Section 6.3.8.

1.2 DIMENSIONAL ANALYSIS

Any equation has to respect the dimensions of the quantities it relates. It makes no sense to say that some number of apples is equal to some number of oranges.

By dimensions we mean the fundamental quantities that are described by the so-called base units of some system. These dimensions are length L, mass M, and time T in the most commonly used systems of units, namely SI and CGS.[2] The SI units of length, mass, and time are the meter (m), kilogram (kg), and second (s), respectively. In CGS, they are centimeter (cm), gram (g), and second (s). Technically, we should also consider temperature, but we can always just multiply it by Boltzmann's constant k and treat it as energy.

[1]Matrices will be discussed in Section 6.3.

[2]When it comes to electricity and magnetism, SI adds a new base unit called the Ampere, while CGS describes charges and current in units derived from mass, length and time. When we are dealing with electromagnetism in this course, I will generally use CGS units.

Quantities that measure length, mass, or time have the dimensions L, M, and T, respectively. Derived quantities have the dimensions of the combination of fundamental quantities from which they are derived. So, for example, velocity has dimensions LT^{-1} and acceleration LT^{-2}. Force comes from mass times acceleration, so the dimensions of force are MLT^{-2}.

I will use square brackets to denote the dimensions of some quantity. So, for example, momentum $p = mv$, so $[p] = MLT^{-1}$. Angular momentum ℓ is the product of a length times momentum, so $[\ell] = ML^2T^{-1}$. Energy E is force times distance so $[E] = ML^2T^{-2}$. Notice that dimensional correctness has to carry over, so if I think of energy instead as $E = mc^2$, then its dimensions are that of mass times the square of velocity, and I get the same result.

If you do a calculation and the dimensionality of your result doesn't make sense, then you had to have made a mistake somewhere! It is always a good idea to check the dimensionality of a calculation.

1.2.1 EXAMPLE: THE RADIUS OF A BLACK HOLE

Let's get the "scale" that determines the radius of a black hole, also known as the "event horizon." If you do a fancy calculation in General Relativity, which is in fact quite difficult, you know that in the end the event horizon R can only depend on the mass m of the black hole, the speed of light c, and Newton's gravitational constant G. You know this because there are no other relevant quantities that enter into the calculation.

So, to within factors of two and π and the like, we write

$$R = G^x m^y c^z \tag{1.2}$$

where the powers x, y, and z can be determined by dimensional analysis. The dimensions of R, m, and c are pretty obvious, namely L, M, and LT^{-1}, respectively. To get the dimensions of G, we can go back to Newton's law of gravity, namely

$$F = G\frac{m_1 m_2}{r^2} \qquad \text{so} \qquad MLT^{-2} = [G]M^2L^{-2}$$

so $[G] = M^{-1}L^3T^{-2}$. This is a handy relation to keep in mind for many problems in dimensional analysis involving gravity.

Now return to (1.2) and write

$$L = M^{-x}L^{3x}T^{-2x}M^yL^zT^{-z} = M^{-x+y}L^{3x+z}T^{-2x-z}$$

This gives us the following three equations:

$$\begin{aligned} -x+y &= 0 \\ 3x+z &= 1 \\ -2x-z &= 0 \end{aligned}$$

Adding the second and third equations gives $x = 1$. The third equation gives $z = -2$, and the first equation gives $y = 1$. Therefore, (1.2) becomes

$$R = Gm/c^2$$

A careful study of General Relativity for a static, spherical gravitating mass m yields the so-called "Schwarzschild radius" $R_S = 2Gm/c^2$ as the distance at which clocks stop. Our simple analysis, though, gets us the scale at which this happens, and in fact the right answer to within a factor of two.

1.3 DERIVATIVE OF A FUNCTION OF REAL VARIABLES

You probably first heard about the derivative of a function graphically, namely the slope of a line tangent to some curve at some point. Physically, the derivative tells you how something *changes*. It's the same thing as the slope of a tangent line, which tells you how fast the curve is changing. In mechanics, you first see the derivative as how fast position changes with time, called the velocity, and how fast velocity changes with time, called the acceleration.

I use generic variables to talk about the concept of a derivative, and also integration. Remember, though, these are just dummy variables. So if I write something like $y = f(x)$ and take the derivative with respect to x, I can just change the names later and talk about $x = f(t)$ to describe the position as a function of time.

1.3.1 FUNCTIONS OF A SINGLE VARIABLE

The derivative a function $y = f(x)$ is defined to be

$$f'(x) = \lim_{\Delta x \to 0} \frac{f(x+\Delta x) - f(x)}{\Delta x} \tag{1.3}$$

Note that the right side of the equation looks like the division of zero by zero. Indeed, the derivative is the ratio of "infinitesimal" quantities. In fact, we often use the notation

$$f'(x) = \frac{dy}{dx}$$

Let's use this definition to find the derivative of $f(x) = x^n$ where n is a positive integer, i.e. $n = 1, 2, 3, \ldots$. Think first about the quantity

$$(x+\Delta x)^n = x^n + nx^{n-1}\Delta x + \cdots$$

where the terms I didn't write all have powers of $(\Delta x)^2$ or greater. Therefore

$$f'(x) = \lim_{\Delta x \to 0} \frac{x^n + nx^{n-1}\Delta x + \cdots - x^n}{\Delta x} = \lim_{\Delta x \to 0} \left(nx^{n-1} + \cdots\right) = nx^{n-1}$$

because the terms I didn't write down all have factors of Δx in them, so they go to zero in the limit.

So, we have proven that $f'(x) = nx^{n-1}$ for $f(x) = x^n$ where n is a positive integer. See the end-of-chapter exercises for a problem to go through some steps and argue that $f'(x) = \alpha x^{\alpha-1}$ where α is any real number. It's also true if α is any complex number, but we'll leave that proof for a later time.

We will stop here for now. We will learn the derivatives of so-called "special functions" when we encounter them in Section 1.5, but we have already gotten very far. For example, it is trivial to prove from (1.3) that, for constants a and b,

$$h'(x) = af'(x) + bg'(x) \qquad \text{where} \qquad h(x) = af(x) + bg(x)$$

which means that we already know how to take the derivative of a polynomial function

$$f(x) = a_0 + a_1 x + a_2 x^2 + \cdots = \sum_{i=0}^{n} a_i x^i \tag{1.4}$$

and we'll see in Section 2.2 that most functions can be expressed this way if we let $n \to \infty$.

1.3.2 DIFFERENTIALS

When we write $f'(x) = dy/dx$ for $y = f(x)$, (1.3) tells us that dy and dx are *infinitesimal* quantities, called "differentials," which can be as small as they need to be. That is, we can ignore them if they are added to anything other than zero, and only the lowest order matters otherwise. For example, something like $dx + dx\,dy + (dx)^2$ is just the same as dx. We say that "only the lowest nonzero order is important."

Differentials are very useful for analyzing physical systems, and we will see this a lot in this course and future physics courses.

Differentials are also a very useful concept for deriving a host of mathematical relations. Consider, for example

$$d(uv) = (u + du)(v + dv) - uv = u\,dv + v\,du$$

for two functions $u = f(x)$ and $v = g(x)$. If we divide through by dx we get

$$\begin{aligned} \frac{d}{dx}(uv) &= u\frac{dv}{dx} + v\frac{du}{dx} \\ \text{or} \qquad h'(x) &= f(x)g'(x) + g(x)f'(x) \end{aligned}$$

for $h(x) = f(x)g(x)$. This is called the "product rule" for differentiation.

Now suppose instead that $h(x) = f(g(x))$, again with $u = f(x)$ and $v = g(x)$. We can derive the so-called "chain rule" for $h'(x)$ by multiplying by "1", that is

$$\begin{aligned} \frac{du}{dx} &= \frac{du}{dx}\frac{dv}{dv} = \frac{du}{dv}\frac{dv}{dx} \\ \text{or} \qquad h'(x) &= f'(g)g'(x) \end{aligned}$$

In other words, take the derivative of the function f with the function g as the argument, and then take the derivative of g with respect to x.

We can illustrate the product rule and chain rule with a simple example, namely

$$f(x) = x^n \qquad \text{and} \qquad g(x) = x^m$$

To check the product rule, let $h(x) = f(x)g(x) = x^{n+m}$. In this case we know that $h'(x) = (n+m)x^{n+m-1}$. The product rule gives

$$h'(x) = nx^{n-1}x^m + mx^n x^{m-1} = (n+m)x^{n+m-1}$$

which is correct. To check the chain rule, use $h(x) = x^{nm}$ in which case $h'(x) = nmx^{nm-1}$, but using the chain rule

$$h'(x) = \left[n\left(x^m\right)^{n-1}\right] mx^{m-1} = nmx^{nm-m+m-1} = nmx^{nm-1}$$

We will see plenty of more interesting examples of the produce rule and chain rule as we go through more mathematics and physics.

1.3.3 HIGHER-ORDER DERIVATIVES

We write the "second derivative" of a function $y = f(x)$ as

$$f''(x) = f^{(2)}(x) = \frac{d}{dx}\frac{dy}{dx} = \frac{d^2y}{dx^2}$$

Higher orders follow naturally, including notation.

It is worth noting that, for some reason, the laws of physics, including Newton's Laws, Maxwell's Equations, the Wave Equation, and many others, only involve first and second derivatives. We'll see more about this in Chapter 3 and beyond.

1.3.4 FUNCTIONS OF TWO OR MORE VARIABLES

For a function $u = f(x,y)$ we define the "partial derivatives"

$$\frac{\partial u}{\partial x} = \lim_{\Delta x \to 0} \frac{f(x+\Delta x, y) - f(x,y)}{\Delta x} \qquad \text{and} \qquad \frac{\partial u}{\partial y} = \lim_{\Delta y \to 0} \frac{f(x, y+\Delta y) - f(x,y)}{\Delta y}$$

That is, we treat $y(x)$ as a constant when we take the partial derivative withe respect to $x(y)$. For more than two variables, for example $u = f(x,y,z)$, the analogy is straightforward.

The product rule and chain rule for partial derivatives follow logically.

We will almost always only be dealing with functions where the partial derivatives commute, that is

$$\frac{\partial^2 u}{\partial x \partial y} = \frac{\partial}{\partial x}\frac{\partial u}{\partial y} = \frac{\partial}{\partial y}\frac{\partial u}{\partial x} = \frac{\partial^2 u}{\partial y \partial x}$$

1.3.5 IMPLICIT DIFFERENTIATION

For some equation of the form $y = f(x)$ you find the derivative dy/dx by calculating $f'(x)$. Suppose, however, that you instead have an equation of the form $f(x,y) = g(x)$, where solving for y is not straightforward. How do you go about finding dy/dx in this case?

Partial differentiation gives you the answer, and a technique called "implicit differentiation." In terms of differentials, $f(x,y) = g(x)$ implies that

$$\frac{\partial f}{\partial x}dx + \frac{\partial f}{\partial y}dy = g'(x)\,dx \qquad \text{or} \qquad \frac{\partial f}{\partial x} + \frac{\partial f}{\partial y}\frac{dy}{dx} = g'(x)$$

which is easily solved for dy/dx, in terms of x and y. This will be perfectly useful for you in many situations.

1.3.6 FINDING MINIMA AND MAXIMA

A very common application of differentiation is to find maximum or minimum values of some function, and the values of the independent variable(s) that give you those extrema. More formally, for $a \le x \le b$, what value of x minimizes or maximizes the function $f(x)$? The generalization to functions of more than one variable is obvious.

It is of course possible that $f(a)$ and/or $f(b)$ are maximum or minimum values. If that's not the case, however, then for some value $x = x_0$ in between, $f'(x_0)$ will be positive (negative) for x slightly smaller than x_0, and negative (positive) if slightly larger, in which case $x = x_0$ is a minimum (maximum). It will often be the case that a function is defined with some "free parameters" and we need to find the derivatives with respect to these parameters and set them to zero. Solving the resulting equations gives us the values of the parameters that we are looking for.

We will see many examples of this kind of thinking, but let's do a simple example where we know what the right answer ought to be. What are the dimensions of a rectangle made from a rope of length L that maximizes the area? Your intuition should tell you that the answer is a square, but let's prove it.

Let the sides of the rectangle be x and y. Then the area is $A = xy$ and to find the maximum we need to set the derivative $dA/dx = 0$. The values of x and y are constrained so that $2x + 2y = L$, so we can solve for $y = (L - 2x)/2$ and

$$A = x\frac{L-2x}{2} = \frac{1}{2}(xL - 2x^2) \qquad \text{so} \qquad \frac{dA}{dx} = \frac{1}{2}(L - 4x) = 0$$

Therefore $x = L/4$ and $y = (L - L/2)/2 = L/4$. Indeed, the answer is a square.

It's slicker to solve this problem with implicit differentiation. We can write

$$\frac{dA}{dx} = y + x\frac{dy}{dx} = 0 \qquad \text{so} \qquad \frac{dy}{dx} = -\frac{y}{x}$$

Differentiating the constraint equation gives

$$2 + 2\frac{dy}{dx} = 0 \qquad \text{so} \qquad \frac{dy}{dx} = -1 = -\frac{y}{x} \qquad \text{and} \qquad y = x$$

and again we see that the answer is a square.

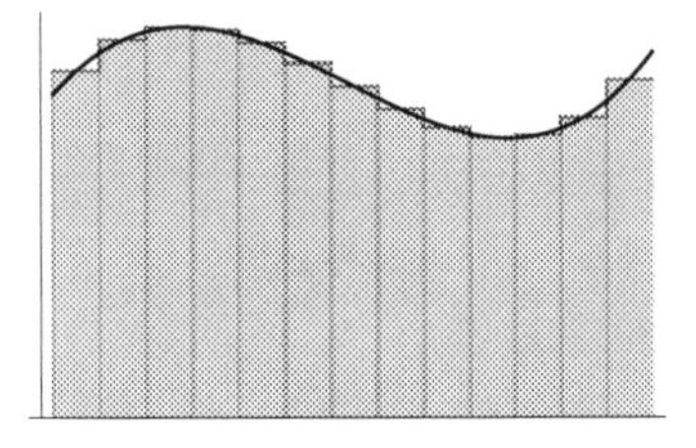

Figure 1.1 The area under a curve, approximated as the sum of a bunch of tall, rectangular boxes. The definite integral is obtained by letting the number N of boxes go to infinity, with the width Δx of each box approaching zero. This gives you a precise result for the area under the curve.

1.4 INTEGRATION

It is tempting to think of integration as the opposite of differentiation, but you should avoid the temptation.

Integration is the sum of a bunch of very small things, technically the infinite sum of infinitesimal things. It is not an accident that the integration sign $\int$ looks like a stretched out S, for "sum."

The reason you might think of integration and differentiation as opposites is because they are connected through the concept of an "antiderivative" and the Fundamental Theorem of Integral Calculus. Our goal in this section is to connect these ideas.

1.4.1 THE DEFINITE INTEGRAL

The "definite" integral is an infinite sum of infinitesimal quantities. (That's the reason the integral sign looks like a tall stretched-out "S," for "Sum.") We start with an ordered set of N real numbers

$$x_i \in \{x_0, x_1, x_2, \ldots, x_N\} \tag{1.5}$$

where $a \leq x_i \leq b$. We then write

$$\int_a^b f(x)\,dx = \lim_{\substack{N\to\infty \\ \Delta x_i \to 0}} \sum_{i=0}^{N} f(x_i)\,\Delta x_i$$

where $x_0 \to a$, $x_N \to b$, and the $\Delta x_i = x_{i+1} - x_i$. The classic picture of the definite integral is "area under a curve." See Figure 1.1. A simple approximation to the area is the sum of large number of narrow vertical rectangular boxes, and the approximation becomes exact in the limits $N \to \infty$ with $\Delta x \to 0$. This is fine and is a nice physical description, but doesn't become useful until we have the fundamental theorem of integral calculus, which we will get to shortly.

When working on some problem in physics, you always want to think of the definite integral as the sum of infinitesimals, defining the infinitesimal quantities in whatever way is handiest for you. Suppose, for example, you want to know the volume of an eggplant, a vegetable with a weird shape. See Figure 1.2. If s is some coordinate that measures the position along the length of the eggplant, then when you slice the eggplant into thin disks, each with thickness ds, then the volume of the

Figure 1.2 To find the volume of an eggplant, you can slice it up perpendicular to its axis. Each slice is a thin disk with volume $\pi r^2 ds$, where $r = r(s)$ is the radius of the particular disk at distance s along the axis. Of course, if you want to find the volume mathematically, you need to know the function $r(s)$ to carry out the integration.

eggplant is

$$V_{\text{eggplant}} = \int_{\text{bottom}}^{\text{top}} \pi r^2 \, ds \qquad \text{with} \qquad r = r(s)$$

Of course, you'll need to know what $r(s)$ is in order to carry out the calculation, but that's something you work out for the particular eggplant you're working on.

As for how you actually carry out the integral mathematically, we need to first discuss the concept of "antiderivative," and then the Fundamental Theorem of Integral Calculus.

1.4.2 THE ANTIDERIVATIVE CONCEPT

I think one reason students often get confused is because we borrow the integration symbol to define the antiderivative as an "indefinite integral." We write

$$\int f(x)dx = F(x)$$

which just means that $f(x) = F'(x) = dF/dx$. However, the reason we use the "integral" symbol for this doesn't really make sense until we connect with the definite integral. We call the function $F(x)$ the "antiderivative" of the function $f(x)$. Sometimes we write

$$\int f(x)dx = F(x) + C$$

which indicates that an arbitrary constant C can always be added to the indefinite integral. We won't usually use this way of writing the antiderivative, but it will come in handy when we are integrating to find the solutions to differential equations.

A more or less obvious example of the antiderivative is

$$\int x^{\alpha}\,dx = \frac{1}{\alpha+1}x^{\alpha+1}$$

This formula breaks down when $\alpha = -1$, however. So, what is the antiderivative of the function $f(x) = 1/x$? We will investigate this in Section 1.5.2.

It is important to note, however, that not all functions have an (analytic) antiderivative. We will encounter these from time to time in physics. Sometimes we will find "tricks" for carrying out the definite integrals nevertheless, but oftentimes the integrals need to be done numerically.

1.4.3 THE FUNDAMENTAL THEOREM OF INTEGRAL CALCULUS

Now we can establish the all-important connection between an integral and an antiderivative. As $N \to \infty$ and the $\Delta x_i \to 0$, the Δx_i become differentials dx, and the sum (1.5) becomes

$$\int_a^b f(x)\,dx = \sum_{i=0}^{\infty} f(x_i)\,dx = \sum_{i=0}^{\infty} \frac{dF(x_i)}{dx}\,dx = \sum_{i=0}^{\infty} \Delta F(x_i)$$

Now the $\Delta F(x_i) = F(x_{i+1}) - F(x_i)$, so each term in this sum cancels with the term before (or the term after), and all that is left are the function $F(x)$ evaluated at the endpoints. That is

$$\int_a^b f(x)\,dx = F(b) - F(a) \equiv F(x)|_a^b$$

This is the result from integral calculus that you undoubtedly have burned into your memory.

Evidently, the sign of the integral changes if the upper and lower limits are interchanged, that is

$$\int_a^b f(x)\,dx = -\int_b^a f(x)\,dx$$

Keep your eyes out for uses of this relationship in this book and elsewhere. It is used a lot in physics.

If the function you're trying to integrate has no analytic antiderivative, then you either find some trick that lets you find the integral (for example, see Section 1.5.6) or you have to resort to numerical integration techniques. Lots of different apps and software libraries provide you the necessary tools.

It can happen often in some fields, statistical mechanics for example, that integration can be used as a technique for approximating difficult sums of finite quantities.

1.4.4 ILLUSTRATION: THE MOMENT OF INERTIA OF A DISK

Let's illustrate these concepts with a simple but important physical problem.

When we describe the motion of a rotating solid object, the concept of "mass" is replaced by "moment of inertia," sometimes called "rotational inertia." Simply put,

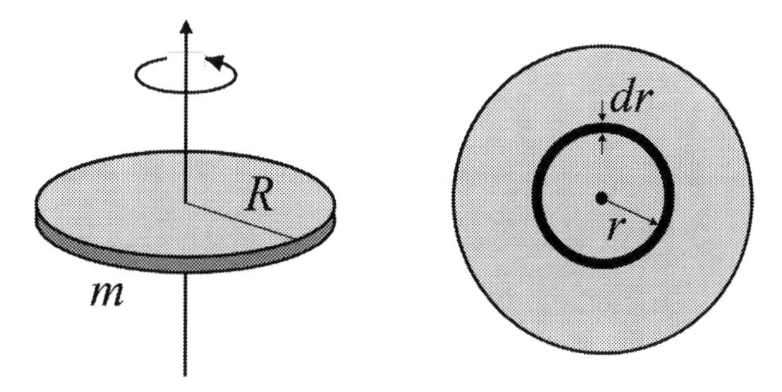

Figure 1.3 Calculating the moment of inertia of a disk of mass m and radius R, by summing up an infinite number of "rings" that each have an infinitesimal thickness dr. You find that the result does not depend on the thickness h of the disk, so it might as well be a cylinder of arbitrary length.

the moment of inertia is the mass times the square of the distance from the axis of rotation. Of course, for a solid object, each atom is, in principle, at a different distance from the axis, so you need to sum up the moments of inertia for all the individual atoms. We might as well treat the atoms as infinitesimal objects, and the sum becomes an integral.

Apply this idea to a solid disk of mass m and radius R that has uniform density. See Figure 1.3. If we divide the disk up into infinitesimally thick rings of radius r, then the moment of inertia is

$$I = \int_{\text{Disk}} dm\, r^2 = \int_0^R dm\, r^2$$

because every piece of an individual ring is at a distance r from the axis. Now dm is just the volume of the ring times the density ρ. Since the ring is infinitesimally thin, we get its volume by multiplying its length $2\pi r$ times the width dr times the disk thickness h. That is

$$dm = \rho \times 2\pi r \times dr \times h = 2\pi\rho h r\, dr$$

The density is just the mass of the disk divided by its volume, so

$$\rho = \frac{m}{\pi R^2 h}$$

Putting all of this together gives the result

$$I = \int_0^R 2\pi \frac{m}{\pi R^2 h} h r\, dr\, r^2 = \frac{2m}{R^2}\int_0^R r^3\, dr = \frac{2m}{R^2}\frac{R^4}{4} = \frac{1}{2}mR^2$$

which is what you'll see in any physics textbook that tabulates the moments of inertia for different solid objects. Notice that the result obviously has dimensions of mass$\times$length^2, as it must.

1.4.5 EVEN AND ODD INTEGRANDS

Consider some integral between equal and opposite sign lower and upper limits, that is

$$I = \int_{-a}^{a} f(x)\,dx \tag{1.6}$$

If $f(x)$ is an *even* function of x, that is $f(-x) = f(x)$, then it is easy to see that

$$\begin{aligned} I &= \int_{-a}^{0} f(x)\,dx + \int_{0}^{a} f(x)\,dx = \int_{a}^{0} f(y)\,(-dy) + \int_{0}^{a} f(x)\,dx \\ &= \int_{0}^{a} f(x)\,dx + \int_{0}^{a} f(x)\,dx = 2\int_{0}^{a} f(x)\,dx \end{aligned} \tag{1.7}$$

(See Section 1.4.6 just below regarding switching variables between x and y.) Pictorially, this is an obvious result. The function $f(x)$ on the positive x-axis is just the mirror image of the function on the negative axis, so the areas under the two sides have to be equal.

Similarly, if $f(x)$ is an *odd* function of x, that is $f(-x) = -f(x)$, the integral (1.6) is zero. In this case, the mirror images are opposite, and the areas under the curves on both sides cancel.

1.4.6 CHANGING VARIABLES IN INTEGRATION

Note that in (1.7), I "changed variables" to $y = -x$ (and then back to $x = y$) in an attempt to show clearly what I was doing. This is not an uncommon technique in integration! Oftentimes it is convenient to express your integral in terms of a different variable from what you start with, or necessary in the case of numerical integration. You can also gain physical insight in many cases by making this so-called "change of variables" in integration. Let's see how this works.

This is, basically, an application of the chain rule in reverse, and we can approach it using differentials. Suppose you are integrating a function $f(x)$ over $a \le x \le b$, but you would prefer to integrate over a variable $y = g(x)$. It will be important that the inverse of the function $g(x)$ be well-defined over $a \le x \le b$, that is $x = g^{-1}(y)$ is known. If we write $c = g^{-1}(a)$ and $d = g^{-1}(b)$ then

$$\int_a^b f(x)\,dx = \int_a^b f(x)\frac{dy}{dy}\,dx = \int_a^b f(x)\frac{dx}{dy}\,dy = \int_c^d f[g^{-1}(y)]\frac{1}{g'[g^{-1}(y)]}\,dy$$

and you are now doing an integral over $c \le y \le d$.

Don't try to use this formula, though. It is easier to just make the change of variables on the specific problem that you are working on. You'll see lots of examples of this after we have defined and worked with the elementary special functions in Section 1.5.

1.4.7 INTEGRATION BY PARTS

If we apply the product rule to integration, then we come up with a technique known as "integration by parts" which can be very useful for solving physical problems and gaining insight to a physical situation. Suppose the integrand $f(x)$ can be split into the product of two functions $u(x)$ and $V(x)$, that is $f(x) = u(x)V(x)$ and you want to carry out the integral

$$\int_a^b f(x)\,dx = \int_a^b u(x)V(x)\,dx$$

Now let the antiderivative of $u(x)$ be $U(x)$, that is $dU/dx = u(x)$, and let $v(x) = dV/dx$ be the derivative of $V(x)$. Then from the product rule

$$\frac{d}{dx}(UV) = \frac{dU}{dx}V + U\frac{dV}{dx} = u(x)V(x) + U(x)v(x)$$

Integrating this equation over $a \leq x \leq b$ and rearranging terms, we get

$$\int_a^b f(x)\,dx = U(x)V(x)|_a^b - \int_a^b U(x)v(x)\,dx$$

and we have "traded" an integral of $u(x)V(x)$ for one of $U(x)v(x)$, which is presumably easier to carry out. It will often happen, for example see Section 7.2, that $U(x)V(x)|_a^b = U(a)V(a) - U(b)V(b)$ vanishes. Furthermore, it won't be uncommon to have things like $a \to -\infty$ and $b \to +\infty$.

1.4.8 INTEGRALS IN HIGHER DIMENSIONS

You probably learned about "double integrals" and "triple integrals" in your calculus classes. These are straightforward generalizations of the one-variable integral, very similar to the generalization to partial derivatives from ordinary derivatives. So, to carry out something of the form

$$\int_a^b dx \int_c^d dy\, f(x,y)$$

you just do the x- (or y-) integral first, and then do the other one. (Notice that I moved around the dx and dy so that I could easily indicate which limits go with which variable.) This form implies that you are doing the integral over the rectangular area delimited by $a \leq x \leq b$ and $c \leq y \leq d$, but of course, that doesn't need to be the case.

More typically, in your physics classes, at least, you will see the form

$$\int_R f(x,y)\,dA$$

where R is some specification of a region in the (x,y) plane over which you are to carry out the integral. We write $dA = dxdy$ for the infinitesimal tile in the (x,y) plane. There are other ways to write this infinitesimal area element, and we'll get to that shortly.

Integrals over volume are also common in physics, so expect to see things like

$$\int_R f(x,y,z)\,dV$$

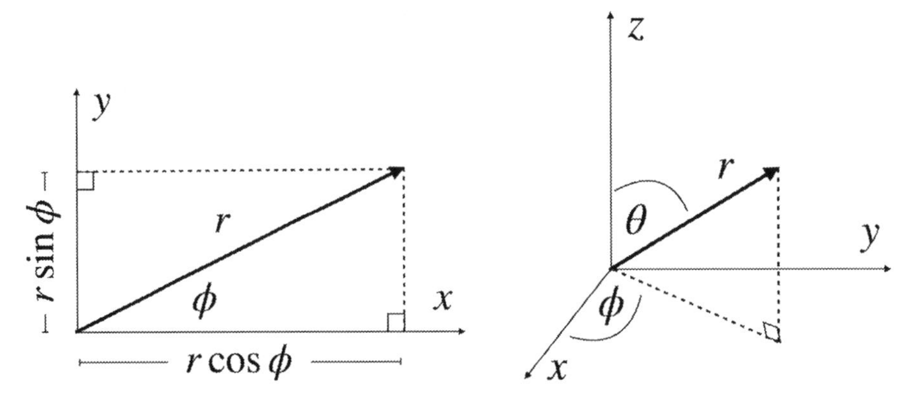

Figure 1.4 Polar (left) and Spherical (right) Coordinate systems with the notation commonly used in physics. We use ϕ to measure the angle of the position vector with respect to the x-axis in both cases. For planar coordinates, the relationships $x = r\cos\phi$ and $y = r\sin\phi$ should be obvious. In spherical coordinates, the length of the projection of the position vector in the xy plane is $r\sin\theta$, where θ is the angle with respect to the z-axis. The relationships for x, y, and z in terms of r, θ, and ϕ follow straightforwardly.

where now R is a region in three-dimensional space, and $dV = dxdydz$. In fact, of course, the integration variables don't need to be real space, and it is not uncommon to encounter things like

$$\int_R f(x_1, x_2, \ldots, x_N)\, dx_1 dx_2 \cdots dx_N$$

and take it as it comes. We'll see specific examples of this sort of thing in this course, and in your more advanced physics courses.

Always remember that an integral is a sum of a bunch of tiny things. You can write that sum in whatever way makes the integral easiest, or do it numerically if you have to.

1.4.8.1 Polar and spherical coordinates

I want to take a little detour here, because this is a good place to introduce the common way that physicists use to work in 2D or 3D space when there is cylindrical or spherical symmetry. I won't formally introduce you to sine and cosine until the next section, but you know enough from high school to follow along what I'm doing here.

We will be discussing different ways of describing vectors in space in Section 4.1, but let's take a moment to use simple geometry to understand how to write the integrand over a planar surface in terms of so-called "polar coordinates" (r, ϕ). That is, we locate a point in space by its distance r from the origin and the angle ϕ it makes with the x-axis. See Figure 1.4. In this case, the infinitesimal area element is

$$dA = (rd\phi) \times dr = r dr d\phi$$

so we would write the integral over some region R of the plane as

$$\int_R f(r,\phi)\, r\,dr\,d\phi$$

This is particularly powerful if the function f depends only on r, and the region R is cylindrically symmetric and goes between two limits r_1 and r_2. In this case

$$\int_R f(r,\phi)\, r dr d\phi = 2\pi \int_{r_1}^{r_2} f(r)\, r\,dr$$

This will often be the case in physical problems of interest.

If we are talking about a region of three-dimensional space which still has the symmetry of a cylinder, then we use the coordinates (r,ϕ,z) and treat the z-coordinate the same as usual.

Perhaps the most common three-dimensional situation in physics is when there is spherical symmetry. In this case, we use the coordinates (r,θ,ϕ), where θ is the polar angle measured down from the z-axis. These coordinates are also illustrated in Figure 1.4. The volume element is

$$dV = (r\sin\theta\, d\phi) \times (r d\theta) \times (dr) = r^2 \sin\theta\, dr d\theta d\phi$$

An often useful change of variables here is $\mu = \cos\theta$, in which case the integral over a spherical volume of radius R (including $R \to \infty$) becomes

$$\int_0^R r^2 dr \int_0^\pi \sin\theta d\theta \int_0^{2\pi} d\phi\, f(r,\theta,\phi) = \int_0^R r^2 dr \int_{-1}^1 d\mu \int_0^{2\pi} d\phi\, f(r,\mu,\phi)$$

where, as we'll see shortly, $d\mu = -\sin\theta\, d\theta$.

Let's illustrate this by finding the volume of a sphere. All we want to do here is add up the volume elements, so we calculate

$$\int_0^R r^2 dr \int_0^\pi \sin\theta d\theta \int_0^{2\pi} d\phi = \int_0^R r^2 dr \int_{-1}^1 d\mu \int_0^{2\pi} d\phi = \frac{R^3}{3} \times 2 \times 2\pi = \frac{4}{3}\pi R^3$$

which is something you probably remember from high school.

1.4.9 INTEGRALS OVER ARBITRARY LINES AND SURFACES

Always remember that an integral is a sum of a bunch of tiny things. I can't say that enough. So, we might write something like

$$\int_C f(x,y,z) d\ell$$

to denote the sum of a bunch of small things $f(x,y,z)d\ell$ along some arbitrary curve C. We'll need more information about how $f(x,y,z)$ varies along C, but that will come with the physics. If the curve C is a closed loop, then we write

$$\oint_C f(x,y,z) d\ell$$

Similarly, we can talk about integrals over some arbitrary surface S. We write

$$\int_S f(x,y,z)\,dA \to \oint_S f(x,y,z)\,dA$$

if the surface is closed. Again, we need to know a lot more about the function and the surface in order to carry out this integral.

Let's illustrate this by finding the surface area of a sphere. The area element dA in spherical coordinates is just the "inside" surface of the volume element at $r = R$, that is $dA = R^2 \sin\theta d\theta d\phi$ or $dA = Rd\mu d\phi$ for $\mu = \cos\theta$. We just sum up the surface area elements to get

$$\int_S dA = R^2 \int_{-1}^{1} d\mu \int_0^{2\pi} d\phi = 4\pi R^2$$

which again should be something you remember from high school.

In Section 4.3 we will learn about some very important theorems that relate the integral around a closed curve to the area that the curve encloses, and the integral around a closed surface to the volume that the surface encloses. Physically, these theorems will tell us about how to learn what's "inside" from what is happening on the "boundaries."

1.5 ELEMENTARY SPECIAL FUNCTIONS

Except for the little detour we took in Sections 1.4.8.1 and 1.4.9, the only functions we have discussed so far are power laws like x^α or their linear combination. Now it's time to more or less precisely define some common "special functions" and to study their properties.

Let me give you a little warning here. With a lot of these special functions, sometimes I will not put the argument in parenthesis, but sometimes I will. For example, sometimes I'll write $\sin(x)$ or $\log(x)$, and sometimes I'll write $\sin x$ or $\log x$. There is not necessarily any rhyme or reason to this. This is not to confuse you on purpose, but to help keep you from getting confused, because different authors write these functions differently.

1.5.1 CIRCULAR FUNCTIONS

I want to clearly define what we mean by the *functions* sine and cosine, and then make connections onto what you learned in your high school trigonometry class. See Figure 1.5.

The cosine and sine functions are defined[3] as functions of the arc length x along a unit circle $u^2 + v^2 = 1$, counter clockwise from the point $(1,0)$ on the (u,v) axes. That

[3]The names "cosine" and "sine" seem reversed to me, also "cotangent" and "tangent." In expressions that involve both $\cos(x)$ and $\sin(x)$, I always write $\cos(x)$ first, as do most people, I think. Maybe there's an interesting story in there somewhere. However, as paraphrased from Shakespeare, "A rose by any other name would smell as sweet."

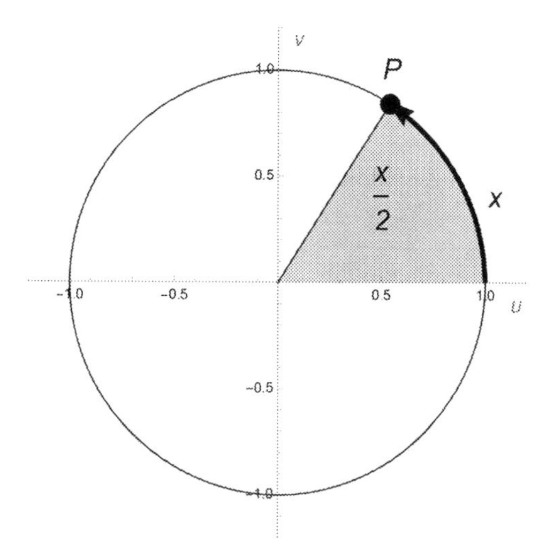

Figure 1.5 Geometry of the unit circle in the (u,v) plane. The variable x measures distance in the direction shown along the circle $u^2+v^2=1$. The coordinates of the point P are defined to be $(\cos x, \sin x)$. The partial area of the circle subtended by the arc is equal to $x/2$. The figure is drawn for $x=1$.

is, we define the coordinates of the point P to be $(\cos x, \sin x)$. Obviously, $\cos^2 x + \sin^2 x = 1$.

The transcendental number π is defined as the ratio of the circumference of a circle to its diameter. Therefore, the circumference of the unit circle is 2π, and if I add 2π to the argument of any circular function, I have to get the same value back. That is

$$\cos(x+2\pi) = \cos(x) \qquad \text{and} \qquad \sin(x+2\pi) = \sin(x)$$

and similarly for the circular functions derived from sine and cosine.

Figure 1.5 shows an alternate way to define the argument x of the circular functions. The area of the "pie slice" is the fraction $x/2\pi$ times the area $\pi(1^2)$ of the unit circle, that is $x/2$. Therefore, we could have defined x to be twice the area of the pie slice, which we would call the area inside the circle that is "subtended" by the arc from $(1,0)$ to the point P. This alternate definition of x will be helpful when we cover the hyperbolic functions in Section 1.5.4.

There are several other functions defined in terms of $\cos(x)$ and $\sin(x)$, namely *tangent* and *cotangent*

$$\tan(x) = \frac{\sin(x)}{\cos(x)} \qquad \text{and} \qquad \cot(x) = \frac{1}{\tan(x)} = \frac{\cos(x)}{\sin(x)}$$

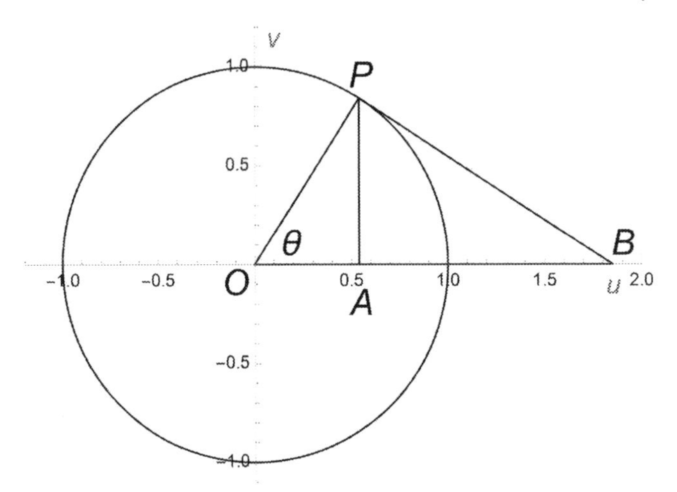

Figure 1.6 This version of Figure 1.5 is useful for comparing the trigonometric definitions of sine and cosine to the functional form. For example, $\cos\theta = OA/OP = \cos(x)/1 = \cos(x)$.

and *secant* and *cosecant*

$$\sec(x) = \frac{1}{\cos(x)} \qquad \text{and} \qquad \csc(x) = \frac{1}{\sin(x)}$$

Figure 1.6 redraws Figure 1.5 using some trigonometry. In high school, you probably learned about $\cos\theta$ and $\sin\theta$ for the right triangle *OPA*. If we want to talk about the angle $\theta \equiv x/R = x$ for our unit circle with radius $R = 1$, then you see that the definitions in terms of "opposite," "adjacent," and "hypotenuse" are exactly the same. Therefore, everything you learned about trigonometry in high school follows from our definition above. For example,

$$\cos\theta = \sin(\pi - \theta) = \frac{OP}{OB} = \frac{1}{OB} \qquad \text{so} \qquad OB = \frac{1}{\cos\theta} = \frac{1}{\cos x}$$

This is a good time to point out that sometimes people quote the argument of circular functions in *degrees* as opposed to *radians*. When we are using the dimensionless arc length of the unit circle, that is x in Figure 1.5, then we are using radians. However, when we are using the angle θ as described in Figure 1.6, sometimes people use degrees ($^\circ$) instead of radians. The circle is divided into 360°, a somewhat arbitrary number that is approximately equal to the number of days in a year, and neatly divisible by lots of numbers, like, 2, 3, 4, 12, and so on. Since there are 2π radians in a circle, in order to convert radians to degrees, you multiply by the factor $360/2\pi = 180/\pi \approx 57.3$. That is, there are about 57.3° in one radian.

There are a bunch of useful trigonometric identities that I'm not going to bother to derive. You can look these up easily enough online, but the following are some of the most useful:

$$\cos(-x) = \cos(x) \tag{1.8a}$$
$$\sin(-x) = -\sin(x) \tag{1.8b}$$
$$\cos(x+y) = \cos x \cos y - \sin x \sin y \tag{1.8c}$$
$$\sin(x+y) = \sin x \cos y + \cos x \sin y \tag{1.8d}$$
$$\cos 2x = \cos^2 x - \sin^2 x \tag{1.8e}$$
$$\sin 2x = 2 \sin x \cos x \tag{1.8f}$$
$$\cos \frac{x}{2} = (\pm)\sqrt{\frac{1+\cos x}{2}} \tag{1.8g}$$
$$\sin \frac{x}{2} = (\pm)\sqrt{\frac{1-\cos x}{2}} \tag{1.8h}$$

You can also get many identities easily with Euler's Formula (Section 2.4), but getting there requires finding the derivatives, which we do next using the identities!

To calculate the derivative of $\cos(x)$, we go back to the fundamental definition of the derivative and make use of the identities and some geometry. We have

$$\begin{aligned}
\frac{d}{dx}\cos x &= \lim_{\Delta x \to 0} \frac{\cos(x+\Delta x) - \cos(x)}{\Delta x} \\
&= \lim_{\Delta x \to 0} \frac{\cos(x)\cos(\Delta x) - \sin(x)\sin(\Delta x) - \cos(x)}{\Delta x} \\
&= -\sin(x) \lim_{\Delta x \to 0} \frac{\sin(\Delta x)}{\Delta x} - \cos(x) \lim_{\Delta x \to 0} \frac{1-\cos \Delta x}{\Delta x}
\end{aligned}$$

The first limit can be evaluated geometrically. Refer to Figure 1.6. The area of the "slice of pie" is $x/2$, and is clearly in between the areas of right triangles *OPA* and *OPB*. It is easy to see that the area of *OPA* is $\cos(x)\sin(x)/2$. If we call $OP = 1$ the base of triangle *OPB*, then the height is

$$PB = \sqrt{OB^2 - OP^2} = \sqrt{\frac{1}{\cos^2 x} - 1} = \frac{\sin x}{\cos x}$$

and the area of triangle *OPB* is $\sin x / 2\cos x$. Therefore

$$\frac{1}{2}\cos x \sin x < \frac{x}{2} < \frac{1}{2}\frac{\sin x}{\cos x} \qquad \text{or} \qquad \cos x \leq \frac{x}{\sin x} < \frac{1}{\cos x}$$

This obviously implies that $\sin x / x \to 1$ as $x \to 0$. Now we also know that

$$\frac{1-\cos x}{x} = \frac{1-\cos x}{x}\frac{1+\cos x}{1+\cos x} = \frac{\sin x}{x}\frac{\sin x}{1+\cos x} \to (1)(0) = 0$$

as $x \to 0$. We have therefore proven that

$$\frac{d}{dx}\cos x = -\sin x$$

Given this result, it is easy to get the derivative of $\sin(x)$ from $\cos^2 x + \sin^2 x = 1$ by taking the derivative with respect to x of both sides. You find

$$\frac{d}{dx}\sin x = \cos x$$

The derivatives of the other circular functions, like $\tan x = \sin x/\cos x$, you can get from the product rule.

I haven't discussed the inverse circular functions, that is $\cos^{-1}(x)$ and $\sin^{-1}(x)$, also known as $\arccos(x)$ and $\arcsin(x)$, but their definition and usage are pretty obvious. There is an important caveat, though, because any one value of $\sin(x)$ or $\cos(x)$ can correspond to many different values of x. For example, $\sin(\pi/4) = 1/\sqrt{2} = \sin(3\pi/4)$, so what is $\sin^{-1}(1/\sqrt{2})$? The answer is that we have a convention that the range of $y = \sin^{-1}(x)$ is $-\pi/2 \leq y \leq +\pi/2$, so $\sin^{-1}(1/\sqrt{2}) = \pi/4$. Similarly, the range of $y = \cos^{-1}(x)$ is $0 \leq y \leq \pi$.

1.5.2 NATURAL LOGARITHMS

I think the best way to present this is the way I learned it fifty years ago, from *Calculus and Analytic Geometry* by Abraham Schwartz.

The derivative of x^n is nx^{n-1}, so the antiderivative of x^n is $x^{n+1}/(n+1)$ which obviously breaks down if $n = -1$. So what is the function for which the derivative is $1/x$? We can suss this out by defining

$$f(x) = \int_1^x \frac{1}{t}dt$$

and show that $f(x)$ has all these properties of a logarithm.[4] For example, it is obvious that, $f(1) = 0$. Next consider, with the change of variables $u = 1/t$,

$$f\left(\frac{1}{a}\right) = \int_1^{1/a} \frac{1}{t}dt = \int_1^a u\left(-\frac{du}{u^2}\right) = -\int_1^a \frac{1}{u}du = -f(a)$$

Similarly prove (a good homework problem) that $f(ab) = f(a) + f(b)$, which implies that $f(a^n) = nf(a)$ for $n \in \mathbb{Z}^+$ (the positive integers).

So hypothesize that $f(x) = \log_b(x)$ and set about to find the base b by going back to the definition to give it the right derivative. That is

$$\begin{aligned}
\frac{d}{dx}\log_b(x) &= \lim_{\Delta x \to 0} \frac{\log_b(x+\Delta x) - \log_b(x)}{\Delta x} = \lim_{\Delta x \to 0} \frac{1}{\Delta x}\log_b\left(1+\frac{\Delta x}{x}\right) \\
&= \frac{1}{x}\lim_{\Delta x \to 0} \frac{x}{\Delta x}\log_b\left(1+\frac{\Delta x}{x}\right) = \frac{1}{x}\lim_{\Delta x \to 0}\log_b\left(1+\frac{\Delta x}{x}\right)^{x/\Delta x} \\
&= \frac{1}{x}\log_b\left[\lim_{n\to\infty}\left(1+\frac{1}{n}\right)^n\right] = \frac{1}{x}
\end{aligned}$$

[4]Maybe we need to review the definition of a logarithm, namely that $\log_b(x)$ is the function for which $b^{\log_b(x)} = x$. It should be clear from this definition that $\log_b(1) = 0$, $\log_b(b) = 1$, $\log_b(xy) = \log_b(x) + \log_b(y)$, $\log_b(1/x) = -\log_b(x)$, and $\log_b(x^\alpha) = \alpha\log_b(x)$.

This determines the value of the logarithm base b, namely

$$b = \lim_{n\to\infty}\left(1+\frac{1}{n}\right)^n \equiv e$$

It is worth taking a few minutes to use a calculator or some app to find the value of $(1+1/n)^n$ for larger and larger values of n. One finds

$$e = 2.7182818\ldots$$

for this new transcendental number.

The function $\log_e(x) \equiv \log(x)$ is called the "natural logarithm." A lot of people write $\ln(x)$ instead of $\log(x)$, but I won't. If I ever need the logarithm to base 10, I will write $\log_{10}(x)$.

To sum things up,

$$\frac{d}{dx}\log(x) = \frac{1}{x} \qquad \text{and} \qquad \int_1^a \frac{dx}{x} = \log(a)$$

1.5.3 EXPONENTIAL FUNCTIONS

The inverse of the natural logarithm function is called *the* exponential function. That is

$$e^{\log(x)} = x$$

which is really just the definition of the logarithm, and

$$\log(e^x) = x\log(e) = x$$

since $\log(e) = 1$, also by definition of the logarithm. An alternative notation for the exponential function is $\exp(x) = e^x$.

So what is the derivative of e^x? We answer this by using the chain rule. Writing $x = \log y$ with $y = e^x$, and taking the derivative of both sides with respect to x, we find

$$1 = \frac{1}{y}\frac{dy}{dx}$$

or $dy/dx = y$. That is

$$\frac{d}{dx}e^x = e^x$$

It is also possible to find the derivative by going back to the definition of e, but that requires an expansion using the binomial theorem, which we really won't get to until Section 9.3.1.

Of course, using the chain rule, putting $y = ax$,

$$\frac{d}{dx}e^{ax} = \frac{de^y}{dy}\frac{dy}{dx} = ae^{ax}$$

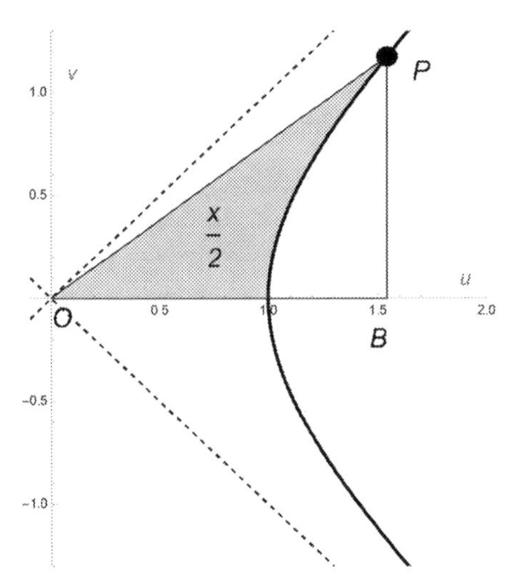

Figure 1.7 Geometry of the unit hyperbola $u^2 - v^2 = 1$ in the (u, v) plane. The variable x measures twice the area subtended by the hyperbolic arc from $(u, v) = (1, 0)$ to the point P at $(u, v) = (\cosh x, \sinh x)$. The figure is drawn for $x = 1$.

Now take a moment to consider the function

$$f(x) = be^{ax} \qquad \text{for which} \qquad f'(x) = abe^{ax} = af(x)$$

In other words, if you are trying to find a function for which, when you take its derivative, you get the same function back but multiplied by a constant a, the answer has to be of the form be^{ax}. This will become a handy way to think about a host of physics problems.

1.5.4 HYPERBOLIC FUNCTIONS

Recall from Section 1.5.1 that the circular functions $\cos x$ and $\sin x$ are defined by the coordinate values along a unit circular arc of length x, and that $x/2$ is the area inside the circle subtended by the arc.

Another pair of functions, called $\cosh x$ and $\sinh x$ are defined analogously along an arc of a unit *hyperbola*. In this case, the definition of x is that $x/2$ is the area subtended by the arc, which is *not* the same as the arc length. See Figure 1.7 and compare to Figures 1.5 and 1.6.

In the (u, v) plane, the unit hyperbola is $u^2 - v^2 = 1$. Therefore, with the point P having coordinates $(\cosh x, \sinh x)$, we must have

$$\cosh^2 x - \sinh^2 x = 1 \tag{1.9}$$

This is similar, but not identical, to the analogous relation for the circular functions.

I will wait until the end of this section to show that x is twice the area subtended by the arc, as shown in Figure 1.7. It's not that hard, but little tedious, and it will be helpful to first have the derivatives of $\cosh x$ and $\sinh x$

It's easy enough to show that $u = \cosh(x)$ and $v = \sinh(x)$ satisfy (1.9) with

$$\cosh(x) = \frac{e^x + e^{-x}}{2} \tag{1.10a}$$

$$\text{and} \qquad \sinh(x) = \frac{e^x - e^{-x}}{2} \tag{1.10b}$$

This also makes it simple to calculate the derivatives of the hyperbolic functions. You find

$$\frac{d}{dx}\cosh x = \sinh x \qquad \text{and} \qquad \frac{d}{dx}\sinh x = \cosh x$$

After we learn Euler's Formula in Section 2.4, we can make an analytic connection to the connection with $\cos(x)$ and $\sin(x)$.

There are lots of interesting parallel relationships between the hyperbolic and circular functions, with sign reversals in many cases. For example,

$$\begin{aligned}\cosh^2 x + \sinh^2 x &= \frac{e^{2x} + 2 + e^{-2x}}{4} + \frac{e^{2x} + -2 + e^{-2x}}{4} \\ &= \frac{e^{2x} + e^{-2x}}{2} = \cosh 2x \end{aligned} \tag{1.11}$$

which is similar to (1.8e). This begs us to try for $\sinh 2x$ as

$$2\sinh x \cosh x = 2\frac{e^x - e^{-x}}{2}\frac{e^x + e^{-x}}{2} = \frac{e^{2x} - e^{-2x}}{2} = \sinh 2x \tag{1.12}$$

and, indeed, the analogy with (1.8f) is borne out.

I didn't bother to make plots of the cosine, sine, logarithmic, and exponential functions because I suspect you know what they look like. However, you may not be familiar with the hyperbolic functions, so Figure 1.8 plots $\sinh x$, $\cosh x$, and also

$$\tanh x = \frac{\sinh x}{\cosh x} = \frac{e^x - e^{-x}}{e^x + e^{-x}}$$

The inverse functions are defined just as you would expect. That is, for example, $y = \cosh^{-1}(x)$ just means that $x = \cosh(y)$. In the case of the inverse hyperbolic tangent, though, we can derive an analytical expression in terms of natural logarithms, and this turns up from time to time in physics problems. Writing $y = \tanh^{-1}(x)$, we have

$$x = \tanh y = \frac{e^y - e^{-y}}{e^y + e^{-y}} = \frac{e^{2y} - 1}{e^{2y} + 1}$$

which can be solved for y to arrive at

$$\tanh^{-1}(x) = \frac{1}{2}\log\left(\frac{1+x}{1-x}\right)$$

Now let's show that $x/2$ is the subtended area in Figure 1.7. This area is the area of the triangle *OBP* minus the area under the hyperbola from $u = 1$ to $u = \cosh x$. We have

$$\text{Area}_{OBP} = \frac{1}{2}(\cosh x)(\sinh x)$$

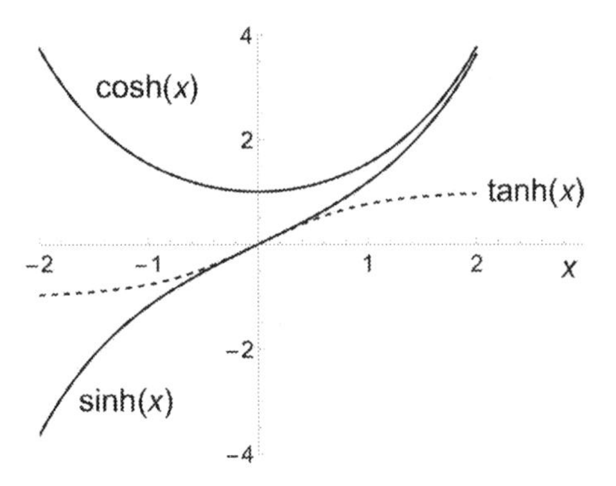

Figure 1.8 Plots of the hyperbolic functions $\cosh(x)$, $\sinh(x)$, and $\tanh(x)$. Note the asymptotic behaviors as $x \to \pm\infty$, in particular that $\tanh(x) \to \pm 1$.

To find the area under the hyperbola, we need to integrate $v(u) = (u^2 - 1)^{1/2}$. We will find the antiderivative of $v(u)$ by using a change of variables to $u = \cosh y$. Then

$$\int (u^2 - 1)^{1/2}\, du = \int \sinh y\,(\sinh y\, dy) = \int \sinh^2 y\, dy$$

Now (1.11) tells us that

$$1 + 2\sinh^2 y = \cosh 2y$$

which lets us carry out the antiderivative as

$$\int \sinh^2 y\, dy = \frac{1}{2}\int (\cosh 2y - 1)\, dy = \frac{1}{4}\sinh 2y - \frac{y}{2} = \frac{1}{2}\sinh y \cosh y - \frac{y}{2}$$

where I've used (1.12). We can now get the area under the hyperbola as

$$\begin{aligned} \text{Area}_{Hyperbola} &= \int_1^{\cosh x} \sqrt{u^2 - 1}\, du \\ &= \int_0^x \sinh^2 y\, dy = \frac{1}{2}\sinh x \cosh x - \frac{x}{2} \end{aligned}$$

Therefore, the area of the subtended arc in Figure 1.7 is

$$\text{Area}_{OBP} - \text{Area}_{Hyperbola} = \frac{x}{2}$$

Finally, I will tell you that there is actually an interesting physical connection between the circular and hyperbolic functions. The circular functions are used to generate rotations in a plane. That is, the "change coordinates" in two dimensional space from some (x, y) to a different (x', y'). In Section 6.5, I will hint at the fact that the hyperbolic functions perform the same "coordinate change" in relativistic "space time," converting some (t, x) to a different (t', x').

1.5.5 THE GAMMA FUNCTION

An important special function that we can now define and work with is

$$\Gamma(z) \equiv \int_0^\infty x^{z-1} e^{-x}\, dx \qquad z \in \mathbb{C} \text{ where } \Re(z) > 0 \tag{1.13}$$

This is a function of a complex number, but in this course we will only refer to the real numbers, in fact only use $z \in \mathbb{Z}^+$, that is the integers $z = n = 1, 2, 3 \ldots$.

The reason physicists find a lot of use for the gamma function is because it is an analytic form for the factorial, which shows up a lot in statistical mechanics and other fields. This is easy to see if we consider integrating by parts. That is

$$\Gamma(n+1) = \int_0^\infty x^n e^{-x}\, dx = \left[x^n(-e^{-x})\right]_0^\infty - \int_0^\infty n x^{n-1}(-e^{-x})\, dx = n\Gamma(n)$$

It is also pretty clear that $\Gamma(1) = 1$ is just from the definition, so repeated application of this result leads us to

$$\Gamma(n+1) = n \cdot (n-1) \cdot (n-2) \cdots (1) = n!$$

for $n \in \mathbb{Z}^+$. Notice that $\Gamma(1) = 1$ tells us that we can consistently define $0! = 1$.

1.5.6 GAUSSIAN INTEGRALS

Physicists make a lot of use of the so-called "error function" in statistical analysis and elsewhere. Refer to Section 9.3.3 for why we see this so often, and where the name comes from. To discuss this, we first need to discuss the concept of the "Gaussian integral."

1.5.6.1 The infinite Gaussian integral

Note that the antiderivative of e^{-x^2} cannot be determined analytically. However, there is a nifty trick we can use to determine the definite integral over $-\infty \leq x \leq \infty$. We write

$$I = \int_{-\infty}^\infty e^{-x^2}\, dx = 2\int_0^\infty e^{-x^2}\, dx$$

where the second relation will come in handy from time to time, and is only the observation that the integrand is an even function of x. Now we obviously can also write

$$I^2 = \left[\int_{-\infty}^\infty e^{-x^2}\, dx\right]\left[\int_{-\infty}^\infty e^{-y^2}\, dy\right] = \int_{-\infty}^\infty \int_{-\infty}^\infty e^{-(x^2+y^2)}\, dxdy$$

If we interepret[5] x and y as variables in the plane, and switch instead to variables r and ϕ, as described in Section 1.4.8.1, we get

$$I^2 = \int_0^{2\pi} d\phi \int_0^\infty r dr e^{-r^2} = 2\pi\left[-\frac{1}{2}e^{-r^2}\right]_0^\infty = \pi$$

[5]We are actually doing a change of variables from (x, y) to (r, ϕ), but I don't want to be too technical here since we haven't gotten yet to the machinery of changing variables in more than one dimension.

where now we realize that the integrand re^{-r^2} *does* have an analytic antiderivative! Therefore $I = \sqrt{\pi}$.

The exponent function in fact usually has a slightly more complicated argument. We write, using a change of variables $y = x\sqrt{a}$ so $dx = dy/\sqrt{a}$, and

$$I(a) = \int_{-\infty}^{\infty} e^{-ax^2} dx = \int_{-\infty}^{\infty} e^{-y^2} \frac{1}{\sqrt{a}} dy = \sqrt{\frac{\pi}{a}} \tag{1.14}$$

1.5.6.2 Derivatives of the infinite Gaussian integral

We can use (1.14) to find other important integrals. For example

$$\int_{-\infty}^{\infty} x^2 e^{-ax^2} dx = -\frac{d}{da} \int_{-\infty}^{\infty} e^{-ax^2} dx = -\frac{d}{da} I(a) = \frac{1}{2}\sqrt{\frac{\pi}{a^3}} \tag{1.15}$$

EXERCISES

1.1 *Given the two complex numbers*

$$z_1 = 1 + i \qquad \text{and} \qquad z_2 = 3 - 4i$$

find the following:

(a) z_1^2
(b) z_1^*
(c) $z_1 z_2^*$
(d) $|z_1|$ *and* $|z_2|$
(e) $|z_1 z_2|$

1.2 *Show that the product of pressure P and volume V has the dimension of energy. Use this and the ideal gas law* $PV = NkT$*, where N is the number of gas molecules and T is temperature in Kelvin, to find the SI units of Boltzmann's constant k.*

1.3 *A simple harmonic oscillator is constructed from a mass m connected to a spring with stiffness k. The stiffness is determined by measuring the force from the spring when it is extended or compressed a certain distance, with the force being proportional to that distance.*

(a) *For classical oscillations with (position) amplitude A, use dimensional analysis to find the energy scale in terms of m, k, and A.*
(b) *For quantum mechanical oscillations, the amplitude is not well-defined, but we expect the energy scale to depend also on* $\hbar$*, which has units of angular momentum. Find the energy scale in terms of m, k, and* $\hbar$*.*

1.4 *A pendulum is made from a mass m hanging from a (massless) string of length* ℓ*. The motion of the string is dictated by the acceleration g due to gravity. Use dimensional analysis to find the relevant time scale* τ *for the pendulum motion. Compare this to what you know to be the period of a pendulum undergoing small amplitude oscillations.*

1.5 *The "Planck Length" ℓ_P is the distance at which gravity is unified with quantum mechanics and relativity. Find an expression for ℓ_P in terms of G, $\hbar$, and c. Evaluate it numerically and compare it to the size of the proton.*

1.6 *The Hubble constant H_0 measures the expansion rate of the universe. It's value has been measured to be 70 km/s per Mpc, where one megaparsec (Mpc) is a common measure of cosmological distances, and equals 3.1×10^{19} km. If H_0 is truly a constant in time, then what is the age of the universe? Express your answer in years.*

1.7 *Use dimensional analysis to estimate the energy of an electron with mass m bound in an atom with size $a = 10^{-10}$ m. In this case, the scale is set quantum mechanically according to the quantity $\hbar$, which has the same dimensions as angular momentum. Express your answer in electron volts. (When working in quantum mechanics, it is handy to remember that $\hbar c = 200$ MeV·fm and that $mc^2 = 0.511$ MeV for an electron.)*

1.8 *Use dimensional analysis to find an expression for the pressure at the center of the Sun, assuming it only depends on gravity and the solar mass and radius. Now assume the Sun has uniform density, is made only of hydrogen, and follows the ideal gas law to find the temperature at the center of the Sun.*

1.9 *In class we showed that the derivative with respect to x for $f(x) = x^n$, where n is a positive integer, is $f'(x) = nx^{n-1}$.*

(a) Show that this relation also holds for $n = 0$.
(b) Use the definition of the derivative to show that this relation also holds when n is a negative integer.
(c) Use (a) and (b) to show that this relation still holds if $f(x) = x^{p/q}$ where p and q are integers, that is, when the exponent is a rational number. Hint: Consider $y^q = x^p$.
(d) Can you use all this to rationalize that the derivative of x^α is $\alpha x^{\alpha-1}$ for any $\alpha \in \mathbb{R}$?

1.10 *For the function $y = \cos^{-1}(x)$, find the derivative dy/dx. I think the simplest way to do this is to make use of differentials.*

1.11 *The equation $ax^2 + by^2 = c$, where a, b, and c are positive constants, describes a collection of points (x,y) that lie on an ellipse. Find the two points at which the slope $dy/dx = 1$ in terms of a, b, and c.*

1.12 *Suppose you have 100 m of fencing and you want to create a rectangular pen for your dog. You have a very long wall that you can use for one side of the pen, and you want to make a pen with the largest area as possible for your dog to run around in. What should you use for the length and width of the pen?*

1.13 *You have a fixed number of square feet of lumber with which to build an open box of maximum volume. The box must have square sides, and no top:*

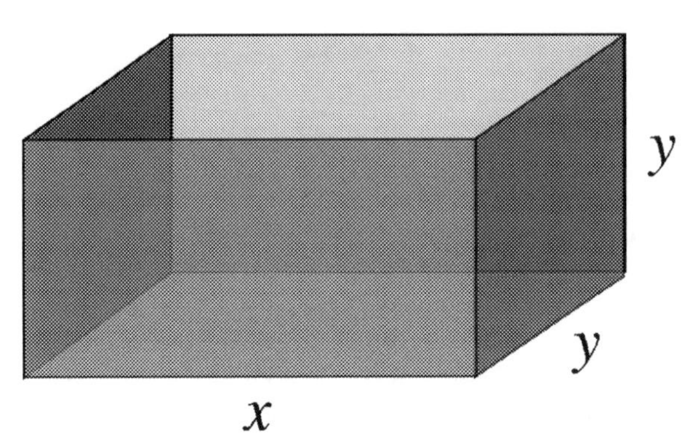

Find the ratio of the length of the base to the height of the box.

1.14 *A lifeguard stands on a beach at point (x_1, y_1) and spots a swimmer in trouble in the water at point (x_2, y_2). She runs on the sand with speed v_1, and swims with a speed v_2. To save the swimmer, she wants to minimize the time it takes to get to him.*

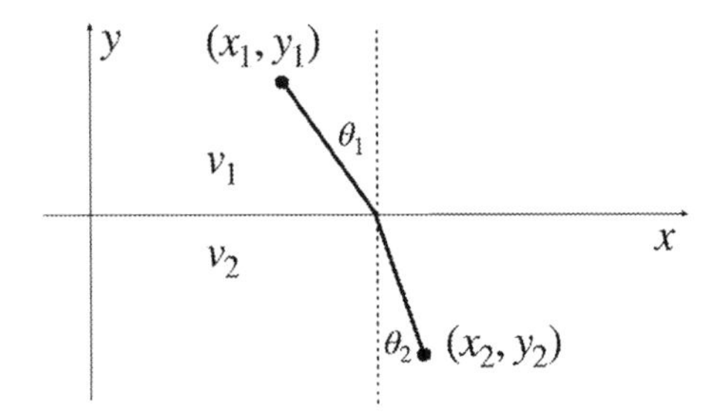

If the coastline coincides with the x-axis above, find the path of least time. Express your answer in terms of the angles θ_1 and θ_2 and the speeds v_1 and v_2. (The result should look familiar to you.)

1.15 *Show that $d(\tan x)/dx = \sec^2 x$ by using the definition of $\tan x$ with the chain rule and the product rule.*

1.16 *Use the change of variables $ax = \tan u$ to evaluate the integral*

$$\int_0^\infty \frac{dx}{1+a^2x^2}$$

1.17 *Using the trigonometric interpretation of the circular functions, show that $\tan\theta$ is the normal to the tangent line of the unit circle at the point given by $(\cos\theta, \sin\theta)$.*

1.18 *Show that the area under the curve $y = \cosh(x)$ over $0 \le x \le a$ is equal to the path length along the arc of the curve between points $(0, 1)$ and $(a, \cosh a)$.*

1.19 *A particle moves in a circle in the (x,y) plane, centered on the origin. Find an expression that relates the velocity $v_x = dx/dt$ in the x-direction and the velocity $v_y = dy/dt$ in the y-direction to the position coordinates x and y. Draw a picture of a circle and indicate a few points on it that convince you that your answer is correct. Of course, you need to explain your reasoning.*

1.20 *The energy of a simple harmonic oscillator made of a mass m and a spring with stiffness constant k imoving in one dimension x is $E = mv^2/2 + kx^2/2$, where $v = dx/dt$.*

- (a) *Take the derivative of the right side, along with Newton's Second Law and Hooke's Law, to show that the energy does not change with time.*
- (b) *Integrate over the quarter of a period where both v and x are positive, and derive an expression for the period T in terms of k and m. The integral is easy to carry out using a change of variables involving a circular function.*

1.21 *A dam in the shape of an inverted triangle blocks a river valley, forming a lake of depth D and width W. Taking the water pressure $p(y) = \rho g y$ at depth y from the surface of the lake, find the total force acting on the dam. Check that your result is dimensionally correct. Calculate the force on the Hoover Dam (W = 200 m) from Lake Mead (D = 160 m). Express your result in tons of force.*

1.22 *Show that $f(x) = \int_1^x (1/t)dt$ has the property $f(ab) = f(a) + f(b)$ using an appropriate change of integration variables. Hence show that $f(a^n) = nf(a)$ for $n \in \mathbb{Z}^+$.*

1.23 *Consider a right circular cone of height h and base radius r, as shown on the right. Let ℓ be the slant height of the cone.*

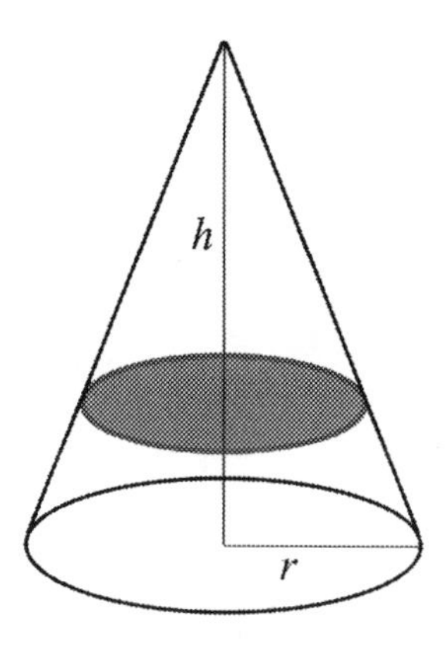

- (a) *Find the volume V in terms of h and r by adding up the volume of a bunch of thin circular disks, one of which is shown shaded.*

(b) *Now find the ratio h/r that maximizes the volume of the cone for a fixed slant length ℓ.*

1.24 *A thin disk of charge q and radius R lies in a plane perpendicular to the z-axis as shown:*

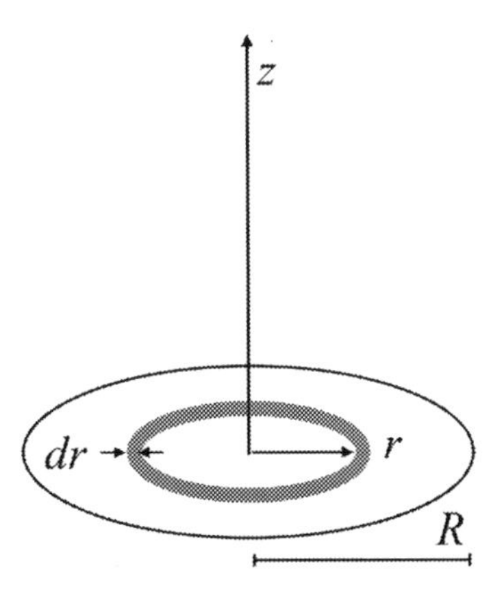

Every point on the charged thin ring of radius r and thickness dr is obviously at the same distance $\rho = \sqrt{r^2 + z^2}$ from any point on the z-axis. The electric potential due to the thin ring at a point z along the axis is therefore $dV = k\,dq/\rho$ where dq is the charge on the thin ring, and $k = 1$ in Gaussian units, or $k = 1/4\pi\varepsilon_0$ in SI units.

(a) *Calculate the potential $V(z)$ from the entire disk. You will likely find it useful to define the planar charge density $\sigma = q/\pi R^2$.*
(b) *Find the electric field $E(z) = -dV/dz$.*
(c) *Find the electric field as $z \to 0$, that is, right up next to the disk. This should agree with the standard problem of the electric field from an infinitely large sheet of charge.*
(d) *Find the electric field as $z \to \infty$ and check that the answer agrees with the field from a point charge.*

1.25 *A flat, square plate of side length L lies in the xy plane, centered on the origin:*

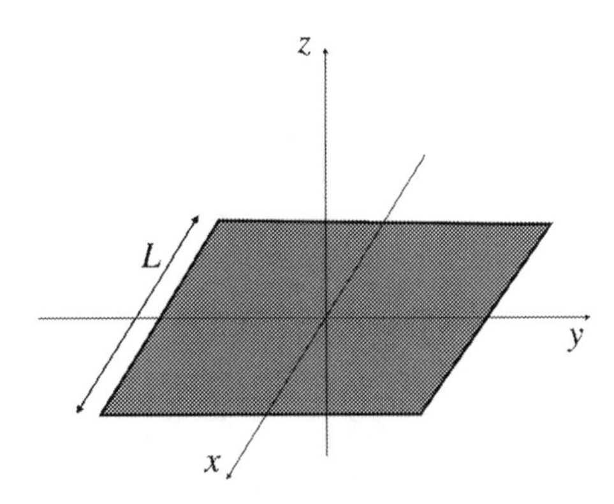

The plate is uniformly charged with total charge Q. Integrate over the surface of the plate to find the electric potential $V(z)$ along the z-axis. Recall that the electric potential at a point a distance r from some charge q is given by

$$V(r) = k\frac{q}{r}$$

where $k = 1$ in CGS units and $k = 1/4\pi\varepsilon_0$ in SI units. Then find the electric field $E(z) = -dV/dz$. Now check that you get the correct answer for the electric field in the limits $z \to 0$ and $z \to \infty$. (You know these answers from your introductory physics course!)

1.26 *The total mechanical energy of a simple pendulum of length ℓ and bob mass m is*

$$E = \frac{1}{2}m\ell^2\dot{\theta}^2 + mg\ell(1-\cos\theta) = mg\ell(1-\cos\theta_0)$$

where $-\theta_0 \le \theta \le \theta_0$ is the angle through which the pendulum swings, and $\dot{\theta} \equiv d\theta/dt$. Find an integral for the period $T(\theta_0)$, divided by the "small angle" period $2\pi\sqrt{\ell/g}$. This is most easily done by integrating from $\theta = 0$ to $\theta = \theta_0$ and then multiplying by four. Calculate the integral numerically using some computer application, and check that your result makes sense in the limits $\theta_0 \to 0$ and $\theta_0 \to \pi$.

1.27 *The motion of a damped harmonic oscillator in one dimension is given by*

$$x(t) = Ae^{-\beta t}\cos(\omega t + \phi)$$

Find A and ϕ in terms of the initial conditions $x(0) = x_0$ and $v(0) = v_0$. Assume that A, β, and ω are all real and positive.

1.28 *Use the definitions of hyperbolic sine and hyperbolic cosine in terms of exponential functions to prove that*

$$\sinh(x+y) = \sinh(x)\cosh(y) + \sinh(y)\cosh(x)$$

1.29 *Evaluate the following integral*

$$\int_0^\infty x^4 e^{-ax^2}\,dx$$

by taking appropriate derivatives of the Gaussian integral. This integral is used to find the root-mean-square velocity of gas particles that follow the Maxell-Boltzmann Distribution in statistical mechanics.

1.30 *Consider a straight rod of length ℓ and mass m. The center of mass of the rod is*

$$x_{\mathrm{CM}} = \frac{1}{m}\int_0^L x\,dm$$

where x measures the position along the rod.

(a) *Show that x_{CM} is what you expect if the rod has uniform mass density.*
(b) *Now calculate x_{CM} assuming that the mass density $\lambda(x)$ of the rod grows linearly from zero at the end of the rod at $x = 0$. Express your answer as a constant times L.*

1.31 *The figure below shows an inverted vertical right circular cone of uniform mass density and height h and base radius R, with symmetry around the z-axis:*

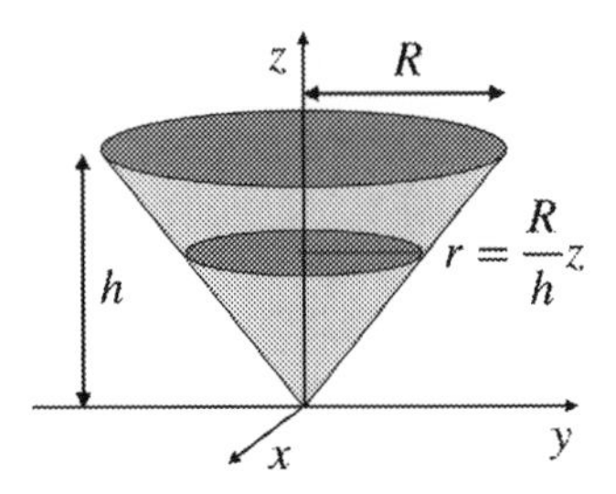

The moment of inertia for an object $\mathcal{O}$ with mass m is given by

$$I = \int_{\mathcal{O}} (x^2 + y^2)\, dm = \int_{\mathcal{O}} \xi^2\, dm$$

where $\xi = (x^2 + y^2)^{1/2}$ is the distance from the z-axis for an infinitesimal mass element dm. Find the moment of inertia of the cone in terms of m, h, and R. You might start by finding the moment of inertia of a disk with radius r and thickness dz.

1.32 *The moment inertia of an object with mass m rotating about a specific axis is $I = \int dm\, r^2$ where r is the distance from the axis of the infinitesimal mass element dm, and the integral is over the entire object. Use this to determine the moment of inertia of a long, thin rod of mass m, length L, and uniform linear mass density λ, rotating about an axis at one end of and perpendicular to the rod, as shown in the figure:*

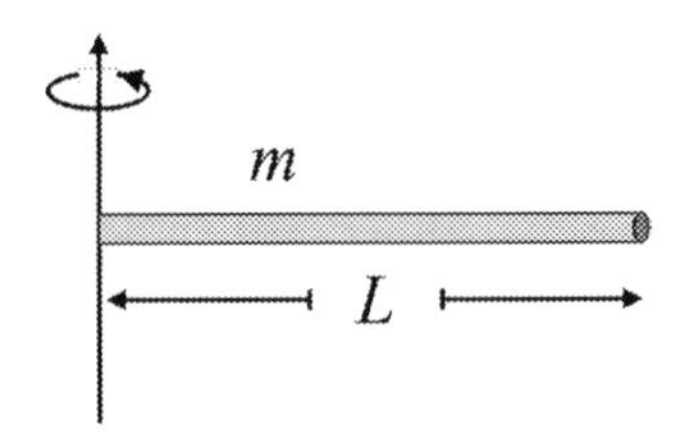

1.33 *A disk of mass M, thickness a, radius R_1, and with a hole of radius R_2 rotates about its symmetry axis, as shown in the figure below. Find the moment of inertia in terms of M, R_1, and R_2.*

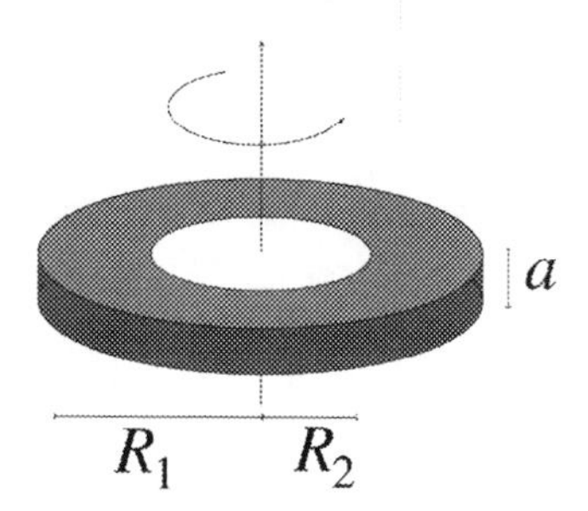

2 Infinite Series

Infinite series appear all of the time in mathematics that is used in physics. They provide excellent ways to connect mathematical concepts and to solve differential equations. Truncated infinite series are also very important in approximations.

You almost certainly covered infinite series and Taylor expansions in your prior calculus courses, so much of this chapter is likely a review. It is presented, though, with a different perspective than your math courses, though, and you might find that helpful. I'm also trying, wherever I can, to give you a physical perspective on the mathematics and to show you how we apply it to some physical problems.

2.1 A SIMPLE POWER SERIES

Let's start with a simple example of a *finite* series, maybe the simplest thing we can think of. We write

$$S_n = 1 + x + x^2 + \cdots + x^n$$

where n is a positive integer. We call this a "power series" in x. There is a nifty trick for finding an analytical expression for S_n. First, multiply this by x, that is

$$xS_n = x + x^2 + x^3 + \cdots + x^{n+1}$$

Subtracting the second expression from the first lets us solve for S_n, namely

$$S_n = \frac{1 - x^{n+1}}{1 - x} \tag{2.1}$$

So what happens if we let $n \to \infty$, that is, let the series become *infinite*? Well, we will get some kind of nonsense if $x \geq 1$. If $x = 1$, then (2.1) gives $0/0$, and if $x > 1$, then it gives ∞. Similar problems come in if we have $x \leq -1$. It seems that to make sense of the case for $n \to \infty$, then, we must require that $-1 < x < 1$, or $|x| < 1$. In this case, we have, for $x \in \mathbb{R}$,

$$\frac{1}{1-x} = 1 + x + x^2 + x^3 + \cdots \qquad \text{where} \qquad |x| < 1 \tag{2.2}$$

This is our first example of an infinite series, and we will see it very often in physics. (In fact, it is a specific case of another example we'll see shortly.) If we put $x \to -x$, then

$$\frac{1}{1+x} = 1 - x + x^2 - x^3 + \cdots \qquad \text{where} \qquad |x| < 1 \tag{2.3}$$

In the case that $|x| \ll 1$, we come up with a very important approximation, namely

$$\frac{1}{1 \pm x} = 1 \mp x + \mathcal{O}(x^2)$$

DOI: 10.1201/9781003355656-2

which we'll use over and over again in physics. The term $\mathcal{O}(x^2)$ means "of order x^2," and is an important notation when we want to estimate how close the approximation is to the truth.

We'll go through these same notions in the rest of this chapter, but in the more general "Taylor Series" which we can apply to functions as a rule.

2.1.1 CONVERGENCE OF THE POWER SERIES

An important mathematical concept, which we will *not* get into in great detail, is the idea of convergence of the infinite series. We saw this a little in our simple example, above, where we argued that (2.2) and (2.3) only made sense if $|x| < 1$. We will delve into this a little bit more deeply in Section 8.1.

For the purposes of this course, don't worry about this too much. You will develop a sense over time if there could be a convergence problem with a series solution to some problem.

2.2 TAYLOR SERIES

Instead of deriving the idea of Taylor Series step by step, I think it is better to give you the answer first, discuss it using some graphics, and then go on to show you how you get there.

Taylor's theorem, which we will (more or less) prove shortly, says that for any function $f(x)$ for which derivatives at a point $x = x_0$ are well-defined, we can write

$$\begin{aligned} f(x) &= f(x_0) + f'(x_0)(x - x_0) + \frac{1}{2!} f''(x_0)(x - x_0)^2 + \cdots \\ &= \sum_{n=0}^{\infty} \frac{1}{n!} f^{(n)}(x_0)(x - x_0)^n \end{aligned} \tag{2.4}$$

where $f^{(n)}(x_0)$ is a shorthand for the nth derivative evaluated at $x = x_0$, that is

$$f^{(n)}(x_0) = \left. \frac{d^n}{dx^n} f(x) \right|_{x=x_0}$$

In other words, pretty much any function can be written as a (possibly) infinite polynomial in the variable $x - x_0$, and this polynomial can be written analytically to any order given an analytic expression for $f(x)$. We will encounter later some "special functions" (like the Bessel function, spherical Bessel function, ...) which only can be written as Taylor series, but let's not dwell on that for now.

It is not necessary for $x \in \mathbb{R}$, but in this course, this is the only case we will study seriously. Expansions of complex functions in terms of complex variables are especially important in many advanced physics courses. We will touch on this more in Chapter 8.

Equation (2.4) may look like a mouthful, but it is actually quite easy to interpret. Look at the first two terms. Ignoring the others gives

$$f'(x_0) = \frac{f(x) - f(x_0)}{x - x_0}$$

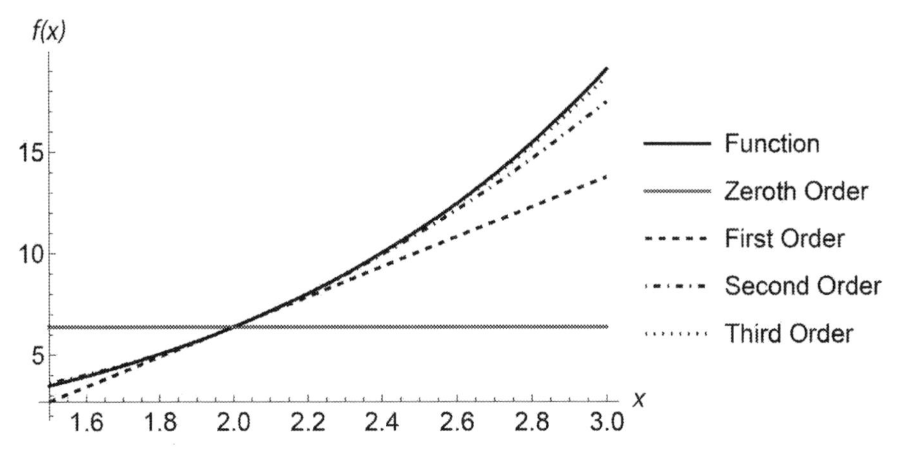

Figure 2.1 Graphic demonstration of a Taylor series expansion. This example expands the function, plotted as a black solid line, about the point at $x_0 = 2$, up to third order.

which says that $f'(x_0)$ is the slope of the straight line that passes through $x = x_0$. In other words, the first two terms are a "straight line" approximation to $f(x)$. Adding higher terms just makes higher-order approximations. Figure 2.1 shows pretty clearly, I think, how the Taylor expansion terms represent a systematically better and better approximation to the function, especially as $|x - x_0| \ll 1$. This is also the idea behind using Taylor expansions as approximations.

Now let's move on to see where (2.4) comes from.

2.2.1 EXPANDING ABOUT $X = 0$

The easiest path to deriving (2.4) is to start with $x_0 = 0$. Assume that we can write

$$f(x) = a_0 + a_1 x + a_2 x^2 + \cdots$$

and see how we might find the coefficients a_n. This is actually pretty simple, since

$$\begin{aligned} f(0) &= a_0 \\ f'(0) &= 1 \cdot a_1 \\ f''(0) &= 2 \cdot 1 \cdot a_2 \\ f'''(0) &= 3 \cdot 2 \cdot 1 \cdot a_3 \end{aligned}$$

and so forth. In other words, $f^{(n)}(0) = n! \cdot a_n$, and so

$$f(x) = f(0) + f'(0)x + \frac{1}{2!} f''(0)x^2 + \cdots = \sum_{n=0}^{\infty} \frac{1}{n!} f^{(n)}(0)x^n \tag{2.5}$$

Remember that $0! = 1$. If you don't believe me, review Section 1.5.5 and find $\Gamma(1)$.

Let's use this approach to derive the Taylor expansion for $f(x) = 1/(1-x)$, that is (2.2). All we need to do is to calculate $f(x)$ and its derivatives at $x = 0$.

$$f(0) = 1 \qquad f'(0) = -(-)\left.\frac{1}{(1-x)^2}\right|_{x=0} = 1 \qquad f''(0) = -2(-)\left.\frac{1}{(1-x)^3}\right|_{x=0} = 2$$

and so forth. In other words, $f^{(n)}(0) = n!$ and we get

$$\begin{aligned} f(x) = \frac{1}{1-x} &= f(0) + f'(0)x + \frac{1}{2!}f''(0)x^2 + \frac{1}{3!}f'''(0)x^3 + \cdots \\ &= \sum_{n=0}^{\infty} \frac{1}{n!}f^{(n)}(0)x^n = \sum_{n=0}^{\infty} x^n \end{aligned}$$

That is, we get (2.2).

2.2.2 SIMPLE EXAMPLES

Let's derive the Taylor expansions (about $x = 0$)[1] for some of the functions from Section 1.5. The simplest is for $f(x) = e^x$, since the derivative of e^x is just e^x so that $f^{(n)}(0) = 1$ for all n, and

$$e^x = 1 + x + \frac{1}{2!}x^2 + \frac{1}{3!}x^3 + \cdots \tag{2.6}$$

This result is used so often in physics that you should memorize it.

The expansions for $\cos(x)$ and $\sin(x)$ are also pretty easy. The derivative of $\cos(x)$ is $-\sin(x)$ and the derivative of $\sin(x)$ is $\cos(x)$. Also, of course, $\cos(0) = 1$ and $\sin(0) = 0$. The series expansions therefore are only the terms that have $\cos(x)$ as the derivative, with alternating signs. You find

$$\cos x = 1 - \frac{1}{2!}x^2 + \frac{1}{4!}x^4 + \cdots \tag{2.7a}$$

$$\sin x = x - \frac{1}{3!}x^3 + \frac{1}{5!}x^5 + \cdots \tag{2.7b}$$

The expansion for $f(x) = \log(1+x)$ is more interesting. In the first place, we consider this instead of $f(x) = \log(x)$ because $f(0)$ would not be defined in this case. The derivatives are straightforward, namely

$$f'(x) = \frac{1}{1+x} \qquad f''(x) = -\frac{1}{(1+x)^2} \qquad f'''(x) = 2\frac{1}{(1+x)^3}$$

which brings us to $f^{(n)}(0) = -(-1)^n(n-1)!$ for $n \geq 1$, and $f(0) = \log 1 = 0$. Therefore

$$\log(1+x) = x - \frac{1}{2}x^2 + \frac{1}{3}x^3 + \cdots \tag{2.8}$$

This is also a formula that you will see often in physics problems.

[1] Taylor expansions about $x = 0$ are sometimes called "Maclaurin expansions."

Another very common expansion in physics is for functions of the form $f(x) = (1+x)^\alpha$, where $\alpha \in \mathbb{R}$. You should take the time to work out the first few terms of the expansion yourself. The result is

$$(1+x)^\alpha = 1 + \alpha x + \frac{1}{2}(\alpha - 1)\alpha x^2 + \frac{1}{6}(\alpha - 2)(\alpha - 1)\alpha x^3 + \cdots \quad (2.9)$$

It should become second nature to you to think that $(1+x)^\alpha = 1 + \alpha x$ when $|x| \ll 1$.

2.2.3 EXPANDING ABOUT $X = X_0$

Now let's return to deriving (2.4). Start with (2.5) and change variables to $y = x + x_0$. Clearly, the derivative with respect to y is the same as the derivative with respect to x, since $dy/dx = 1$. Setting $x = 0$ is the same as setting $y = y_0$. Therefore, (2.5) becomes

$$f(y) = \sum_{n=0}^{\infty} \frac{1}{n!} f^{(n)}(x_0)(y - x_0)^n$$

Then just write $y \to x$ and you recover (2.4) and we're done.

As an example, let's expand $f(y) = \log(y)$ about $y = 1$ and compare to (2.8). Since

$$f'(y) = \frac{1}{y} \qquad f''(y) = -\frac{1}{y^2} \qquad f'''(y) = 2\frac{1}{y^3}$$

and so on, we have $f(1) = 0$, $f'(1) = -1$, $f''(1) = 2$, etc.... Therefore we end up with

$$\log(y) = (y-1) - \frac{1}{2}(y-1)^2 + \frac{1}{3}(y-1)^3 + \cdots$$

Switching to $x = y - 1$ recovers (2.8).

2.3 EXPANSIONS AS APPROXIMATIONS

Physicists make a lot of use of (2.4) and (2.5) in approximations. So long as $|x| \ll 1$ in (2.5), or $|x - x_0| \ll 1$ in (2.4), the successive terms in the expansion will be smaller and smaller as the order increases. It is therefore likely that the first term or two or three will be plenty good enough for your application.

We refer to approximation schemes like this as "expanding in some small parameter." Oftentimes, though, the quantity x, or $x - x_0$, will have dimensions, so we need to know that it is small compared to something else? *The best approach here is always to express your formula in terms of a dimensionless parameter before expanding.*

We can illustrate this with a cute physics problem. We refer to the acceleration of gravity near the Earth's surface as g, a number likely ingrained in your memory as $g = 9.8\ \mathrm{m/s^2}$. We treat g as a constant for the motion of falling bodies, projectiles, and the like. However, since the force of gravity decreases as we move out from the center of the Earth, we expect g to actually be a function of the height y above the Earth's surface. The dependence on y should be very weak, since any change in height would be much smaller than the Earth's radius R, but it could be important if we are making a precise measurement of the motion.

So how can we get a useful approximation for $g(y)$ that is good for $y \ll R$? The force of gravity on an object of mass m that is at a distance $r = R + y$ from the center of the Earth is

$$F_{\text{grav}} = G\frac{mM}{r^2} = G\frac{mM}{(R+y)^2} = mg$$

where M is the mass of the Earth. Therefore

$$g(y) = \frac{GM}{(R+y)^2} = \frac{GM}{R^2}\left(1+\frac{y}{R}\right)^{-2}$$

This formula is "exact" (for a spherical Earth of uniform density), but its form makes it difficult to solve differential equations for the motion of objects. On the other hand, I have written this formula in terms of a small parameter $y/R \equiv x$, and can make use of the expansion (2.9) to simplify it. We have

$$g(y) \approx \frac{GM}{R^2}\left(1-2\frac{y}{R}\right) = g_0(1-\beta y)$$

where $g_0 = gM/R^2 = 9.8\ \text{m/s}^2$ and $\beta = 2/R = 3.1\times10^{-7}/\text{m}$. In other words, near the Earth's surface, the acceleration due to gravity decreases linearly with height, by a fractional amount of 3.1×10^{-5} per 100 m.

2.3.1 POTENTIAL ENERGY MINIMUM

There is one series expansion that is often used as an approximation, because it reduces a problem that is generally intractable to one that is very familiar, namely simple harmonic oscillation.

If a potential energy function $U(x)$ has a minimum at $x = x_0$, then the derivative there is zero and

$$U(x) = U(x_0) + \frac{1}{2}\left.\frac{d^2U}{dx^2}\right|_{x_0}(x-x_0)^2 + \cdots$$

See Figure 2.2. This means that the motion of a particle whose energy confines it to a region close to the minimum will undergo approximate simple harmonic motion with an effective "spring constant" $k = d^2U/dx^2$ evaluated at $x = x_0$.

This result has very many practical uses, because most systems in nature have a minimum in their potential energy function that defines the stable configuration. So-called "small oscillations" about this minimum have many physical consequences. One that you actually might observe in a classroom laboratory has to do with the quantum mechanical energy levels of the nitrogen molecule N_2. In this case, the potential energy function between the two nitrogen atoms is complicated, but the energy levels corresponding to vibrational motion are very close to that predicted for a harmonic oscillator.

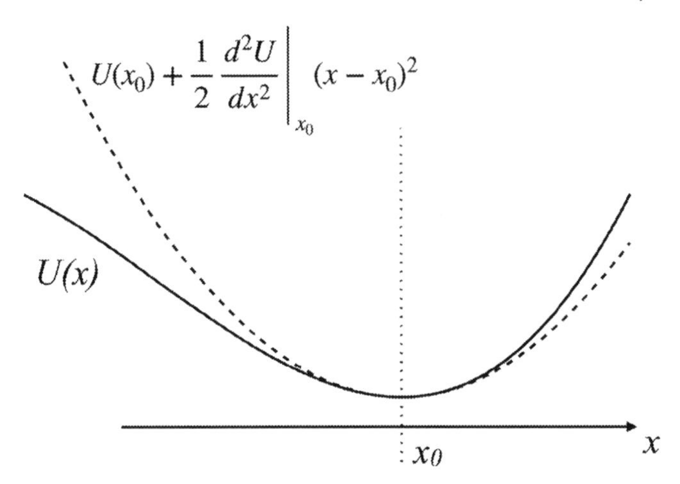

Figure 2.2 This figure illustrates how a potential energy function can be approximated by a Taylor series expansion about a local minimum. The resulting motion is that of a simple harmonic oscillator for a particle that stays close to the minimum.

2.4 EULER'S FORMULA

We are now in a position to derive one of the most important and useful formulas in physics. You will use this result, called "Euler's Formula," in almost every physics course you take.

Consider the expansion about $x = 0$ for e^x given by (2.6). Now instead of expanding the function e^x, expand the function $f(x) = e^{ix}$. Since $i^2 = -1$, we know that $i^3 = -i$, $i^4 = 1$, $i^5 = i$, and so on. That is, $i^n = \pm i$ for n odd, and $i^n = \pm 1$ for n even. This gives

$$\begin{aligned} e^{ix} &= 1 + ix - \frac{1}{2!}x^2 - i\frac{1}{3!}x^3 + i^4\frac{1}{4!}x^4 + i\frac{1}{5!}x^5 + \cdots \\ &= 1 - \frac{1}{2!}x^2 + \frac{1}{4!}x^4 + \cdots \\ &\quad + i\left(x - \frac{1}{3!}x^3 + \frac{1}{5!}x^5 + \cdots\right) \end{aligned}$$

We recognize the two expansions above from (2.7) and arrive at Euler's Formula:

$$e^{ix} = \cos x + i\sin x \tag{2.10}$$

One of the first places we use Euler's Formula is in oscillations, where instead of writing $\cos\omega t$ and $\sin\omega t$ we'll find it much handier to write $e^{\pm i\omega t}$.

Turning Euler's Formula around gives us

$$\cos x = \frac{e^{ix} + e^{-ix}}{2} \tag{2.11a}$$

and

$$\sin x = \frac{e^{ix} - e^{-ix}}{2i} \tag{2.11b}$$

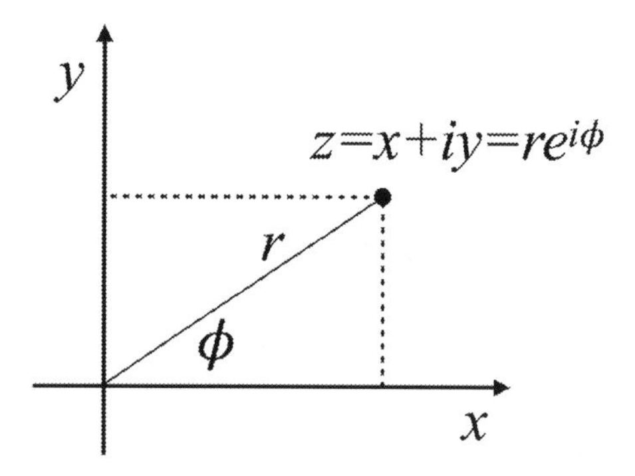

Figure 2.3 Polar representation of a complex number.

which bear a striking resemblance to (1.10). In fact, we see that

$$\cosh(ix) = \cos(x) \qquad \text{and} \qquad \sinh(ix) = i\sin(x)$$

2.4.1 TRIGONOMETRIC IDENTITIES

Euler's Formula makes it easy to find various trigonometric identities. For example, writing $e^{i(x+y)} = e^{ix}e^{iy}$ becomes

$$\begin{aligned}\cos(x+y) + i\sin(x+y) &= (\cos x + i\sin x)(\cos y + i\sin y) \\ &= \cos x\cos y - \sin x\sin y + i(\sin x\cos y + \cos x\sin y)\end{aligned}$$

Just equate real and imaginary parts to get (1.8c) and (1.8d).

I don't think it is fair to call these "derivations" because in order to get Euler's Formula, I needed to know how to find the derivatives of $\cos x$ and $\sin x$, and I got these by using the trigonometric identity for the cosine of the sum. Nevertheless, this is a useful technique for finding more identities, or at least figuring them out if you need them and can't look them up easily.

2.4.2 POLAR REPRESENTATION OF COMPLEX NUMBERS

Euler's Formula gives us a very useful way to express complex numbers in a polar form.

See Figure 2.3. The point in the complex plane can be identified either by its Cartesian coordinates (x, y), or by its polar coordinates (r, ϕ). That is, we can write $z = x + iy$ as $z = re^{i\phi} = r\cos\phi + ir\sin\phi$ since $x = r\cos\phi$ and $y = r\sin\phi$.

We call $r = (x^2 + y^2)^{1/2}$ as the *amplitude* and $\phi = \tan^{-1}(y/x)$ the *phase* of the complex number z. This terminology gains physical importance when we apply it to the amplitude and phase of an oscillating system. See Section 3.4.

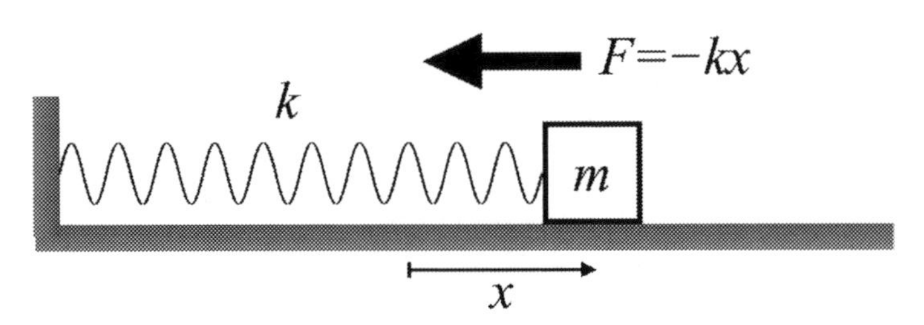

Figure 2.4 The physical situation leading to simple harmonic oscillation. The variable $x = x(t)$ measures the position of the mass m relative to the equilibrium point at $x = 0$. The "restoring force" $F = -kx$ wants to pull the mass back to the equilibrium position.

2.4.3 APPLICATION: SIMPLE HARMONIC MOTION

This is a good time to make use of Euler's formula to solve a basic and crucially important problem in elementary mechanics, namely Simple Harmonic Motion. It is also an opportunity to introduce the material on Ordinary Differential Equations that is the focus of Chapter 3.

Figure 2.4 shows the physical situation. A mass m is attached to a spring with stiffness k, and can slide without friction on a horizontal surface. There is a restoring force $F = -kx$ acting on the mass, where x measures the distance relative to the equilibrium[2] point. The (horizontal) acceleration of the mass is $\ddot{x}(t) = d^2x/dt^2$. Newton's Second Law "$F = ma$" tell us that the force must equal the mass times the acceleration, so

$$-kx = m\frac{d^2x}{dt^2} \qquad \text{or} \qquad \frac{d^2x}{dt^2} = -\omega_0^2 x(t) \qquad \text{where} \qquad \omega_0^2 \equiv \frac{k}{m} \tag{2.12}$$

Equation (2.12) is a *differential equation* that would be solved for the motion $x(t)$, that is, the function $x(t)$ that maps time onto position.

Euler's formula gives us an elegant way to find the solution $x(t)$ to (2.12). In words, we need to find a function that, when you take its derivative *twice*, returns the same function but multiplied by $-\omega_0^2$. You know that the derivative with respect to x of a function $f(x) = e^{ax}$ is just $f'(x) = ae^{ax}$. Therefore, the derivative with respect to t of $e^{i\omega_0 t}$ is $i\omega_0 e^{i\omega_0 t}$, and taking the derivative a second time just brings down another factor of $i\omega_0$. That is

$$\frac{d^2}{dt^2}e^{i\omega_0 t} = (i\omega_0)^2 e^{i\omega_0 t} = -\omega_0^2 e^{i\omega_0 t}$$

which is just what we wanted!

However, there is also a second choice, namely

$$\frac{d^2}{dt^2}e^{-i\omega_0 t} = (-i\omega_0)^2 e^{-i\omega_0 t} = -\omega_0^2 e^{-i\omega_0 t}$$

[2]We are being a little cavalier about what we mean by "equilibrium point." This will be taken up more carefully in Section 3.4.1.

So, how do we deal with the fact that there are two different solutions $x(t)$ to the differential equation?

In fact, we'll find in Chapter 3 that any differential equation with a second derivative will end up having two possible solutions. We'll also learn that, in fact, there are an *infinite* number of solutions for this kind of differential equation, called a *linear* differential equation because the function $x(t)$ and its derivatives only appear in linear form. There are an infinite number of solutions because any linear combination of the two different "independent" solutions is also a solution. That is, the general solution to (2.12) is

$$x(t) = Ae^{i\omega_0 t} + Be^{-i\omega_0 t} \tag{2.13}$$

where A and B are arbitrary constants. We can see this explicitly just by taking the second derivative. That is,

$$\frac{d^2}{dt^2}(A^{i\omega_0 t} + Be^{-i\omega_0 t}) = -\omega_0^2 A^{i\omega_0 t} - \omega_0^2 Be^{-i\omega_0 t} = -\omega_0^2(A^{i\omega_0 t} + Be^{-i\omega_0 t})$$

Now our question is how do we find the constants A and B. For this, we put in the physics of the so-called "initial conditions." Where is the mass m at $t = 0$, and how fast, and in what direction, is it going at this point? Let's say the initial position is x_0. Then

$$x(0) = A + B = x_0$$

We get a second equation for A and B by considering the velocity, that is

$$v(t) = \dot{x}(t) = \frac{dx}{dt} = i\omega_0 A^{i\omega_0 t} - i\omega_0 Be^{-i\omega_0 t}$$

If the initial velocity is v_0, then

$$v(0) = i\omega_0 A - i\omega_0 B = v_0 \qquad \text{or} \qquad A - B = \frac{v_0}{i\omega_0}$$

We now have two equations to solve for A and B in terms of the physical initial values of position and velocity, x_0 and v_0. The equations are simple to solve. You find

$$A = \frac{1}{2}\left(x_0 - i\frac{v_0}{\omega_0}\right) \qquad \text{and} \qquad B = \frac{1}{2}\left(x_0 + i\frac{v_0}{\omega_0}\right)$$

(Notice that $B = A^*$. This will become handy shortly.) We can now write down the solution $x(t)$ that satisfies the differential equation and also the initial conditions:

$$\begin{aligned} x(t) &= Ae^{i\omega_0 t} + Be^{-i\omega_0 t} \\ &= \frac{1}{2}\left(x_0 - i\frac{v_0}{\omega_0}\right)(\cos\omega_0 t + \sin\omega_0 t) + \frac{1}{2}\left(x_0 + i\frac{v_0}{\omega_0}\right)(\cos\omega_0 t - \sin\omega_0 t) \\ &= x_0\cos\omega_0 t + \frac{v_0}{\omega_0}\sin\omega_0 t \end{aligned} \tag{2.14}$$

(Notice how the i magically disappears, but it had to; $x(t)$ is a real function with real initial conditions.) You should convince yourself that this solves the differential equation (2.12), and satisfies the initial conditions.

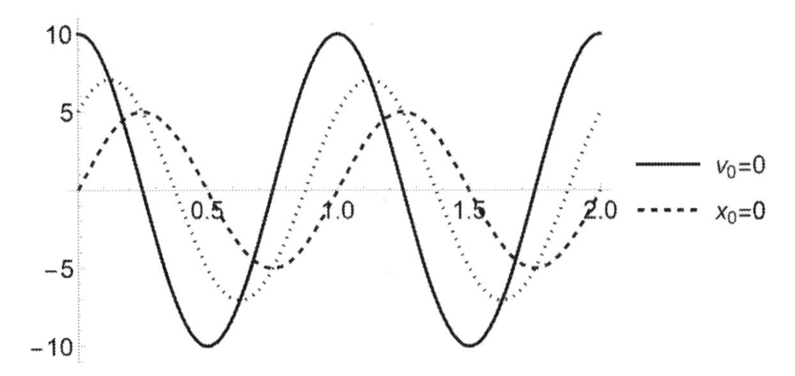

Figure 2.5 Simple harmonic motion with different combinations of initial position and velocity, corresponding to different amplitudes and phases. The solid curve is for zero initial velocity, the dashed curve is for zero initial position, and the dotted curve is when neither of the two are zero. See the text for details.

We can also use Euler's formula to write the solution in an alternate form, which gives some physical insight into what's going on. Since $B = A^*$, we can write

$$A = \frac{1}{2}Re^{i\phi} \qquad \text{and} \qquad B = \frac{1}{2}Re^{-i\phi}$$

where R and ϕ are related to the initial conditions through

$$R = \left(x_0^2 + \frac{v_0^2}{\omega_0^2}\right)^{1/2} \qquad \text{and} \qquad \phi = -\tan^{-1}\left(\frac{v_0}{\omega_0 x_0}\right) \tag{2.15}$$

In this case, we write our solution as

$$\begin{aligned} x(t) &= Ae^{i\omega_0 t} + Be^{-i\omega_0 t} \\ &= \frac{1}{2}R^{i\phi}e^{i\omega_0 t} + \frac{1}{2}Re^{-i\phi}e^{-i\omega_0 t} \\ &= \frac{1}{2}R\left[e^{i(\omega_0 t+\phi)} + e^{-i(\omega_0 t+\phi)}\right] \\ &= R\cos(\omega_0 t + \phi) \end{aligned} \tag{2.16}$$

We call R as the "amplitude" of the oscillation; it is the maximum value that $x(t)$ can reach, so the mass m moves between $+R$ and $-R$. The (angular) frequency of the oscillation is ω_0, so the period is $T = 2\pi/\omega_0$. The phase ϕ measures the "time lag" of the oscillation. If $v_0 = 0$, then $\phi = 0$ and the mass starts at the maximum value; it's as if you pull the mass and let it go (from rest) at $t = 0$. If the initial position $x_0 = 0$, however, then $\phi = -\pi/2$, and the motion starts a quarter period "behind."

Figure 2.5 shows three curves of $x(t)$ for three different choices of initial conditions. Each curve is drawn for $\omega_0 = 2\pi$, that is oscillation period $T = 1$. The solid curve uses $x_0 = 10$ and $v_0 = 0$, while the dashed curve is plotted for $x_0 = 0$ and

$v_0 = 5\omega_0$. Notice that the dashed curve lags the solid curve by a quarter of a period, as expected given that their relative phases are $\pi/2$. The dotted curve corresponds to $x_0 = 5$ and $v_0 = 5\omega_0$, and appropriately lags by a phase of $\pi/4$.

2.5 EXPANSIONS IN MORE THAN ONE VARIABLE

From time to time, it will be useful to perform Taylor expansions in functions of more than one variable. The generalization is straightforward. Up to second order, for a function of two variables, you find

$$\begin{aligned} f(x,y) &= f(x_0,y_0) + \left.\frac{\partial f}{\partial x}\right|_{x=x_0,y=y_0}(x-x_0) + \left.\frac{\partial f}{\partial y}\right|_{x=x_0,y=y_0}(y-y_0) \\ &+ \frac{1}{2!}\left.\frac{\partial^2 f}{\partial x^2}\right|_{x=x_0,y=y_0}(x-x_0)^2 \\ &+ \left.\frac{\partial f}{\partial x}\frac{\partial f}{\partial y}\right|_{x=x_0,y=y_0}(x-x_0)(y-y_0) \\ &+ \frac{1}{2!}\left.\frac{\partial^2 f}{\partial y^2}\right|_{x=x_0,y=y_0}(y-y_0)^2 + \cdots \end{aligned} \tag{2.17}$$

There are neater ways to write this once we have matrix formalism under our belt.

EXERCISES

2.1 *Find the first three nonzero terms of the Taylor expansions about* $x = 0$ *for* $f(x) = \cosh(x)$ *and* $f(x) = \sinh(x)$. *Make sketches of each of these two functions along with the approximations based on the first, second, and third terms.*

2.2 *Using an appropriate trigonometric change of variables, show that*

$$\tan^{-1}(x) = \int_0^x \frac{du}{1+u^2}$$

and then use this to show that

$$\frac{\pi}{4} = 1 - \frac{1}{3} + \frac{1}{5} - \frac{1}{7} + \cdots$$

2.3 *Make a plot of the function* $\log(1+x)$ *over the range* $-1 < x \leq 1.5$. *Then add to the plot successive approximations by powers of* x *for* $x_0 = 0$. *That is, make something similar to Figure 2.1.*

2.4 *The "effective potential energy" for a planet orbiting a star can be written as*

$$U(r) = -\frac{2a}{r} + \frac{b}{r^2}$$

where a *and* b *are positive constants. Find the radius* $r = r_0$ *that gives the minimum value of* $U(r)$, *in terms of* a *and* b. *(This would be the radius of a circular orbit.) Now you know that the angular frequency for a mass m in a potential energy function* $U(x) = kx^2/2$ *is* $\omega = \sqrt{k/m}$. *Use this, and a Taylor expansion of* $U(r)$ *about the minimum value, to find the approximate period of small oscillations of the planet of mass m about* $r = r_0$.

2.5 *Express* $\cos^2(x/2)$ *in terms of* $\cos x$ *by first writing* $\cos(x/2)$ *using Euler's Formula. Use your result to derive Equation (1.8g).*

2.6 *Derive expressions for* $\cos(3x)$ *and* $\sin(3x)$ *in terms of* $\cos x$ *and* $\sin x$ *by applying Euler's Formula.*

2.7 *Two electric charges* $\pm q$ *lie at* $z = \pm a/2$ *on the z-axis.*

(a) *Find the magnitude of the electric field on the z-axis at distances far from the origin. Express your result in terms of the* electric dipole moment $p = qa$. *Compare how the field from an electric dipole falls with distance with that of an isolated electric charge.*
(b) *Repeat for a position on the x-axis, again, far from the origin. Indicate the direction of the electric field relative to that in (a).*

2.8 *The energy of a relativistic particle of (rest) mass m is*

$$E = \sqrt{p^2c^2 + m^2c^4}$$

where p is the momentum and c is the speed of light. Find an expression for E when $p \ll mc$ *to lowest nonzero order in p and physically interpret the terms in the expression.*

2.9 *Consider the function* $f(x) = x^n e^{-x}$. *Find the value of x which maximizes* $f(x)$, *and sketch the function for some large value of n. Then write* $x = e^{\log x}$ *and write* $f(x)$ *in terms of* $y \equiv x - n$. *Expand the logarithm to second order in a Taylor series about* $y = 0$ *and show that* $f(x)$ *is a constant times a Gaussian function of y. Use this result, along with the definition of the Gamma function and Gaussian integrals to derive Stirling's Approximation, namely*

$$n! \approx \sqrt{2\pi n}\left(\frac{n}{e}\right)^n \qquad \text{for} \qquad n \gg 1$$

2.10 *Reduce the following complex expressions into a simple complex (or purely real or purely imaginary) number of the form* $z = x + iy$:

(a) i^i
(b) $\left[(1+i\sqrt{3})/(\sqrt{2}+i\sqrt{2})\right]^{50}$
(c) $\sinh(1 + i\pi/2)$
(d) $e^{2\tanh^{-1} i}$

3 Ordinary Differential Equations

This chapter is about "ordinary" differential equations, that is, equations whose solutions are functions of a single independent variable. We will cover "partial" differential equations in Section 4.5.

We should take a moment to review the concepts of "independent" and "dependent" variables. Consider the relation $y = f(x)$ for the two variables x and y, related by the function f. We call x the *independent variable* because you can, in principle, set it to whatever value you like. On the other hand, y is a *dependent variable* because its value depends on x and what the function $f(x)$ does to x.

Remember that this course is about mathematics as applied to problems in physics. There are many subtleties in the mathematics of differential equations that we will gloss over or outright ignore. For this reason, I encourage all physics students to take a mathematics course in differential equations. Oftentimes, a course is offered which covers differential equations as well as the fundamentals of linear algebra.

An excellent textbook is *Elementary Differential Equations and Boundary Value Problems*, by Boyce, DiPrima, and Meade, now in its 11th edition, published by Wiley. The ISBN is 978-1-119-32063-0. It's the book I learned the subject from (in its 2nd edition) and updates over the years have kept up with education research and the greater available of computing capability for numerical solutions.

3.1 DIFFERENTIAL EQUATIONS AND THEIR SOLUTIONS

Differential equations are equations that involve derivatives of some function. A solution to a differential equation is the function itself. For each order of derivative in the differential equation, you have to specify some "boundary condition" to go along with all of the lower-order derivatives and for the function. (In physics, when the differential equation involves functions of, and derivatives with respect to, time, we call the boundary conditions "initial conditions.") Sometimes you will have a set of differential equations involving different functions, that can appear in some or all of the equations. This is called a system of "coupled" differential equations.

It's worth pointing out at the start that physics is generally formulated in terms of differential equations. Newton's Second Law $\vec{F} = m\vec{a}$, Maxwell's Equations, and the Schrödinger Equation are all differential equations. Einstein's General Relativity gravitational field equations are also differential equations, although highly nonlinear and coupled, so very difficult to solve.

There are very many techniques for solving the very many different types of differential equations. We'll see a small number of these techniques in this course,

DOI: 10.1201/9781003355656-3

but it is often the case that you can just use your wits to figure out a solution. As with most things, practice makes perfect.

In broad terms, there are two general classes of differential equations, called *Ordinary Differential Equations* (ODE's) and *Partial Differential Equations* (PDE's). An ODE has only one independent variable, which we will usually (but not always!) call x or t, and the derivatives of the solution function will only be ordinary derivatives with respect to this variable. The solution to a PDE will be a function of two or more variables, with various partial derivatives with respect to them.

The techniques for dealing with ODE's are very different than those for working with PDE's, so we discuss them separately. We take up a discussion of PDE's n Section 4.5.

3.1.1 ORDINARY DIFFERENTIAL EQUATIONS

An ordinary differential equation for a function $y = y(x)$ is an equation of the form

$$\frac{d^n y}{dx^n} = f\left(x, y, \frac{dy}{dx}, \frac{d^2 y}{dx^2}, \dots, \frac{d^{n-1} y}{dx^{n-1}}\right) \tag{3.1a}$$

where f is some function, and $d^n y/dx^n$ is the nth derivative of $y(x)$. We will often use the notation y' for dy/dx, y'' for d^2y/dx^2, and $y^{(n)}(x)$ for $d^n y/dx^n$. Any solution to (3.1a) must also solve some given boundary conditions at a point $x = x_0$, namely

$$y(x_0) = y_0, \qquad y'(x_0) = y_0', \qquad y''(x_0) = y_0'', \dots \qquad y^{(n-1)}(x_0) = y_0^{(n-1)} \tag{3.1b}$$

Beware of the economy of notation! Writing $y(x)$ just means that the dependent variable y is a function of x. (Mathematicians do not like this convention.) Also, be aware that $y(x_0)$ means the function $y(x)$ evaluated at $x = x_0$, but y_0 is just some number, as is x_0.

Physics problems in classical mechanics, which make use of Newton's Second Law, result in differential equations of some position variable as a function of time. If x measures the position, then the function we're solving for would be $x(t)$, where t is time. That is, x is the dependent variable and t is the independent variable. In this case, we refer to "initial conditions" instead of boundary conditions. We also use a notation, which I think was invented by Newton, where a "dot" over a dependent variable means "time derivative." In other words

$$\dot{x}(t) \equiv \frac{dx}{dt} \qquad \text{or} \qquad \ddot{x}(t) \equiv \frac{d^2 x}{dt^2}$$

For example, we would write Newton's Second Law "$F = ma$" as the differential equation $F(x,t) = m\ddot{x}$ for motion in one dimension x. In this book, I will move back and forth between these two notations, mostly so that you get used to it.

We call n the "order" of the differential equation, that is, the highest derivative that appears. An ODE of order n needs to have n boundary conditions. You can think of each of these conditions leading to determine the constant of integration that we'll get from integrating the derivatives.

If (3.1a) can be cast into the form

$$a_0(x)y + a_1(x)y' + a_2(x)y'' + \cdots + a_n(x)y^{(n)} = g(x) \tag{3.2}$$

then we call the differential equation "linear." If the function $g(x) = 0$ we call the equation "homogeneous." There are some general approaches we can take for linear, homogeneous equations, and these can be directly applied to linear inhomogeneous equations. Let's see how this works in general now. We will return to this in many examples later.

All linear, homogeneous differential equations have an extremely important property that is inherent in so many physical situations. This is the *principle of superposition* which says that any linear combination of two solutions is also a solution. Writing $y(x) = c_1y_1(x) + c_2y_2(x)$, where $y_1(x)$ and $y_2(x)$ are both solutions to (3.2) and c_1 and c_2 are constants, then it is a simple matter to show that $y(x)$ is also a solution. For an nth order equation, we can expect to find n different *linearly independent* solutions so that

$$y(x) = c_1y_1(x) + c_2y_2(x) + \cdots + c_ny_n(x) \tag{3.3}$$

is the general solution. In principle, we can determine the constants $c_1, c_2, \ldots, c_n$ by applying the boundary conditions.

How do we know that the solutions $y_i(x)$ are linearly independent? There is a straightforward way to answer this using matrix manipulations.[1] We apply this solution to the boundary conditions and get a set of equations that can be written as

$$\begin{bmatrix} y_1(x_0) & y_2(x_0) & \cdots & y_n(x_0) \\ y_1'(x_0) & y_2'(x_0) & \cdots & y_n'(x_0) \\ \vdots & \vdots & \cdots & \vdots \\ y_1^{(n)}(x_0) & y_2^{(n)}(x_0) & \cdots & y_n^{(n)}(x_0) \end{bmatrix} \begin{bmatrix} c_1 \\ c_2 \\ \vdots \\ c_n \end{bmatrix} = \begin{bmatrix} y_0 \\ y_0' \\ \vdots \\ y_0^{(n)} \end{bmatrix} \tag{3.4}$$

This set of equations only has a solution if the determinant of the matrix is nonzero, and this is the condition we need for the set of equations $y_i(x)$ to be linearly independent. This determinant is called the *Wronskian* and we write

$$W(x) = \begin{vmatrix} y_1(x) & y_2(x) & \cdots & y_n(x) \\ y_1'(x) & y_2'(x) & \cdots & y_n'(x) \\ \vdots & \vdots & \cdots & \vdots \\ y_1^{(n)}(x) & y_2^{(n)}(x) & \cdots & y_n^{(n)}(x) \end{vmatrix} \tag{3.5}$$

So, linear independence at a point $x = x_0$ requires that we have $W(x_0) \neq 0$. In most of the cases you'll see in physics, $W(x)$ will be nonzero for the entire relevant range of the independent variable x.

This is all fine for homogeneous linear equations, but what if the equation is inhomogeneous? In this case, your first job is to fine a "particular" solution $y_P(x)$ which

[1] We won't be discussing matrices until Chapter 6, so if you're unfamiliar with the way (3.4) and (3.5) are written, just skip over this discussion. I'll be illustrating with examples later in this chapter.

solves (3.2) for nonzero $g(x)$. This solution does not need to have any constants that fit the boundary conditions, because once you have $y_P(x)$, you can add to it any solution (3.3) to the homogeneous equation and the result still solves (3.2). You then apply the boundary conditions to determine the constants c_i.

Again, we will see examples of all this in this chapter.

Finally, we point out that a nonlinear differential equation can also be homogeneous, but in this case you would have to cast it into a form using a change of variables so that it has no remaining explicit function of x with no factor that depends on y or its derivatives. For a first-order equation, for example, a homogenous differential equation is one of the form

$$\frac{dy}{dx} = G\left(\frac{y}{x}\right)$$

I don't think you'll ever have occasion in your upper level physics courses to deal with inhomogeneous nonlinear differential equations.

3.1.2 EXISTENCE AND UNIQUENESS

It might bother you that I have just gone ahead willy nilly writing down some properties of differential equations and implying that I can find solutions with these ideas. But maybe there are other solutions to the equation and boundary conditions that can't be found by following these properties? Or maybe there is no solution to the equation after all?

In fact, the answer is "No" to both questions. For all of the cases we'll study in this course, any given ODE or PDE, along with a set of boundary conditions, has exactly one solution. That is, the solution *exists* and is *unique*. I'm not going to prove this, but you'll do so in a real course in differential equations, and see the conditions that need to be satisfied for the proof.

Nevertheless, this is a very powerful statement. We can go ahead and find a solution to a differential equation, and the boundary conditions, any way that we want, and be assured that it is the right answer. My favorite example of this is the "image charge" approach to solving boundary value problems in electrostatics, something you'll encounter in your Electricity and Magnetism course.

3.1.3 USING SCALED VARIABLES

We have already learned about dimensional analysis. This will allow you to identify fixed quantities in your differential equation that have dimensions of, say, distance, time, energy,..., and then define new variables that are dimensionless.

If your problem is one that requires you to make approximations as to whether something is "big" or "small," then scaling the variables lets you decide this based on whether something is much larger or much smaller than unity. Also, if your problem is suited to solving numerically, then it will be very handy to express your independent and dependent variables in dimensionless form by dividing them by some appropriate scale.

We'll see lots of examples of this, but here's a simple one. The differential equation for the simple harmonic oscillator in one dimension $x(t)$, from $F = ma$ written as $ma - F = 0$, is

$$m\frac{d^2x}{dt^2} + kx = 0$$

The frequency scale is $\omega = (k/m)^{1/2}$. (You should confirm for yourself that this has dimensions of inverse time!) Defining a dimensionless time $\tau = \omega t$ turns the equation into

$$\frac{d^2x}{d\tau^2} + x = 0$$

and now you have a natural way to discuss the oscillator in terms of short times, i.e. $\tau \ll 1$, or long times, i.e. $\tau \gg 1$. (This is actually more useful when the oscillator is also subjected to a damping force and oscillating forcing function, each of which introduce their own time scales.)

3.2 FIRST-ORDER EQUATIONS

First we'll make some general comments and observations about first-order equations, showing how special cases can be solved straightforwardly. Then we will do a couple of examples. Remember, though, that sometimes the best way to solve the equation is to look at it and see your way through without having to go to a menu of techniques!

In the mathematics below, I will make a lot of use of generic functions like $p(x)$ or $Q(x, y)$ and so forth. Do not confused these with being anything other than an arbitrary function used to make a point in a specific discussion. Remember that variable names don't mean anything until I tell you they do, and here I'm only using them to make a point.

3.2.1 SEPARABLE EQUATIONS

If a differential equation has the form

$$\frac{dy}{dx} = \frac{p(x)}{q(y)} \qquad \text{i.e.} \qquad q(y)dy = p(x)dx$$

then it is called *separable*. If the antiderivatives $Q(y)$ and $P(x)$, of $q(y)$ and $p(x)$, respectively, are known, then just integrate both sides to get

$$Q(y) = P(x) + C$$

where the constant of integration C is determined from the boundary condition. In principle, this equation can be solved for the function $y = y(x)$, but depending on the physical problem you're working on, this might not be necessary.

A simple and useful way to apply the boundary condition $y(x_0) = y_0$ is to use it directly in the integration step. Changing the variable names to "primed" version to avoid confusion, we have

$$\int_{y_0}^{y} q(y')\,dy' = \int_{x_0}^{x} p(x')\,dx' \qquad \text{so} \qquad Q(y) - Q(y_0) = P(x) - P(x_0)$$

which can be presumably solved for $y(x)$, or $x(y)$, depending on what the physical situation needs.

3.2.2 EXACT EQUATIONS AND INTEGRATING FACTORS

Imagine a function $\psi(x,y)$, a function of the two independent variables x and y. Its differential is

$$d\psi = \frac{\partial \psi}{\partial x}dx + \frac{\partial \psi}{\partial y}dy$$

Now suppose we have an equation of the form

$$P(x,y)dx + Q(x,y)dy = g(x)$$

If the left hand side was equal to some differential $d\psi$, then integrating the left side is simply $\psi(x,y)$, and the right hand side is just some integral over x. The equation could then be solved, in principle, for the function $y(x)$.

Such differential equations are called *exact*. That is, they must satisfy

$$P(x,y) = \frac{\partial \psi}{\partial x} \qquad \text{and} \qquad Q(x,y) = \frac{\partial \psi}{\partial y}$$

which, by the reversibility of the order of partial derivatives, clearly implies that we must have

$$\frac{\partial P}{\partial y} = \frac{\partial Q}{\partial x}$$

This is very specific and it is unlikely to be such an equation! However, oftentimes it is possible to multiply the equation through by an "integrating factor" $\mu(x)$ which in fact renders the equation exact.

Let's apply this idea to the inhomogeneous linear first-order equation

$$\frac{dy}{dx} + p(x)y = g(x) \tag{3.6}$$

Multiply through by $\mu(x)$ and do a little rearranging to get

$$\mu(x)p(x)y\,dx + \mu(x)\,dy = \mu(x)g(x)\,dx$$

In order for this equation to be exact, we must have

$$\frac{\partial}{\partial y}[\mu(x)p(x)y] = \frac{\partial}{\partial x}[\mu(x)] \qquad \text{or} \qquad \mu(x)p(x) = \mu'(x)$$

which means that

$$\log \mu(x) = \int p(x)\,dx \qquad \text{or} \qquad \mu(x) = \exp\left[\int p(x)dx\right]$$

Indeed, it is fairly obvious that (3.6) becomes

$$\frac{d}{dx}\left[\mu(x)y(x)\right] = \mu(x)g(x)$$

and so the general solution is

$$y(x) = \frac{1}{\mu(x)}\left[\int \mu(x)g(x)\,dx + C\right]$$

where, once again, we determine C from whatever is the boundary condition.

3.2.3 EXAMPLE: RADIOACTIVE DECAY

Quantum mechanics predicts that the probability of radioactive decay for an unstable nucleus to decay in any given short time period dt is λdt, where λ is a constant that can be calculated, in principle, from the properties of the nucleus. Suppose that you start with a given number N_0 of nuclei. Then you expect the number $N(t)$ to decrease over time because they decay. Given the probability λdt, the change in the number of nuclei during this time period is $dN = -\lambda N\,dt$.

This all leads to the differential equation and initial condition

$$\frac{dN}{dt} = -\lambda N(t) \qquad N(0) = N_0$$

The form of this equation makes it simple to guess the answer. With the derivative proportional to the function, you know that the solution has to be some exponential. Let's use a technique from Section 3.2 to get the solution, though.

This is a perfect example of a separable equation. We write

$$\frac{dN}{N} = -\lambda\,dt \qquad \text{so} \qquad \int \frac{dN}{N} = -\lambda \int dt \qquad \text{or} \qquad \log N = -\lambda t + C$$

Exponentiating both sides give us $N = e^{-\lambda t}e^{C}$. Applying the boundary condition gives $N_0 = e^{C}$. Therefore, the solution is

$$N(t) = N_0 e^{-\lambda t}$$

Typically, we tabulate the *half-life* $t_{1/2}$ of radioactive nuclei, namely the time it takes for the sample to decay to one half its value. That is

$$\frac{1}{2} = e^{-\lambda t_{1/2}} \qquad \text{so} \qquad t_{1/2} = \frac{1}{\lambda}\log 2$$

This law also governs the decay of elementary particles and excited atomic states. For some bizarre historical reason, however, we do not quote the half-life for these decays. Instead, we use the mean life $\tau \equiv 1/\lambda$.

3.2.4 EXAMPLE: FALLING UNDER DRAG

This is a simple problem that is intuitively easy to visualize, and is a problem that can be solved using either separability or with an integrating factor.

A mass m falls vertically, starting from rest at height h. In addition to gravity (near the Earth's surface) there is a drag force proportional to the velocity. Find the velocity as a function of time. Then find the vertical position as a function of time. (This involves writing down and solving a first-order equation, and then another first-order equation.)

Let $v(t)$ be the velocity and $y(t)$ be the vertical position as a function of time. Write the drag force as $-bv$ where $b > 0$. (The sign ensures that the drag force is opposite to the direction of velocity.) Assume that positive v is "up." Then

$$m\frac{dv}{dt} = -mg - bv \qquad \text{so} \qquad \frac{dv}{dt} + \frac{b}{m}v = -g$$

which is exactly the form given by (3.6). So, let's approach the solution using an integrating factor. We have

$$\mu(t) = \exp\left[\int \frac{b}{m}\,dt\right] = e^{bt/m}$$

The solution is then

$$\begin{aligned} v(t) &= \frac{1}{e^{bt/m}}\left[\int e^{bt/m}(-g)\,dt + C\right] \\ &= e^{-bt/m}\left[-\frac{mg}{b}e^{bt/m} + C\right] = -\frac{mg}{b}\left[1 - \frac{b}{mg}Ce^{-bt/m}\right] \end{aligned}$$

The initial condition is $v(0) = 0$ so $C = mg/b$ and we finally have

$$v(t) = -\frac{mg}{b}\left[1 - e^{-bt/m}\right]$$

It is always important to check that the result makes sense. Firstly, $v(0) = 0$, as we required. Secondly, $v(t) < 0$ for $t > 0$, which is also correct, since the mass falls from rest. Thirdly, as you might have expected, a "terminal velocity" $v_{\text{term}} = mg/b$ is reached for long times. You expect this because if the mass is moving fast enough, then the force of gravity mg is exactly balanced by the drag force bv.

We can also do some simple dimensional analysis to check the result. For example, for the exponential argument to make sense, the quantity b/m must have the dimensions of inverse time. Since bv is a force, we must have

$$[b][v] = [b]LT^{-1} = MLT^{-2} \qquad \text{so} \qquad [b] = MT^{-1}$$

and, indeed, b/m has dimensions of inverse time.

Now, in this case, the result for $v(t) = dy/dt$ can easily be integrated to find $y(t)$ subject to the initial condition $y(0) = h$. This gives the position as a function of time.

This is a cheaters way of solving the second-order equation for $y(t)$, and that's fine, but not always an option.

Finally, note that we could also have approached this as a separable equation. We have

$$\frac{dv}{1+bv/mg} = -g\,dt \qquad \text{so} \qquad \int_0^v \frac{dv}{1+bv/mg} = -g\int_0^t dt$$

where we integrate both sides, the left from 0 to v and the right from 0 to t. (We can change the integration variables to "primes" if that makes people feel better, but you want to get used to not having to do that.) Then

$$\frac{mg}{b}\log\left(1+\frac{b}{mg}v\right) = -gt \qquad \text{so} \qquad v(t) = -\frac{mg}{b}\left(1-e^{-bt/m}\right)$$

3.2.5 EXAMPLE: THE ROCKET EQUATION

A nice physical example of finding and solving a differential equation for motion in one dimension is the rocket.[2] A rocket accelerates by spewing some of its mass backward out the rear, propelling it forward. The motion is peculiar because the rocket mass decreases over time, but we can analyze the motion using the techniques of differential equations.

The guiding principle here is conservation of momentum $p = mv$, namely

$$\dot{p} = dp/dt = 0$$

if no external forces act on an object with mass m and velocity v. Now imagine a rocket, burning fuel and therefore decreasing its mass while increasing its velocity. If v_{ex} is the exhaust velocity, relative to the rocket, the momentum of the rocket plus the ejected fuel of mass dm at a time $t+dt$ is

$$p(t+dt) = (m+dm)(v+dv) - dm(v-v_{\text{ex}}) = mv + mdv + v_{\text{ex}}dm$$

Note that I neglected the second-order term $(dm)(dv)$, which is allowed because we are working with infinitesimal quantities. Now the change in momentum in time dt is

$$dp = p(t+dt) - p(t) = p(t+dt) - mv = mdv + v_{\text{ex}}dm = 0$$

which is a separable first-order differential equation. With a little rearranging, we can write

$$v_{\text{ex}}\frac{dm}{m} = -dv \qquad \text{so} \qquad v_{\text{ex}}\log\frac{m}{m_0} = -v + v_0$$

where $v = v_0$ is the initial velocity and m_0 is the initial mass. This gives the velocity as a function of mass as

$$v = v_0 + v_{\text{ex}}\log(m_0/m)$$

[2]See Taylor *Classical Mechanics* Section 3.2 for a more detailed discussion.

The velocity as a function of time is determined by the rate at which mass is consumed, so rocket engineers make use of this equation to design space launches, including "multi stage" rockets which shed structures as the fuel they hold is consumed, giving large reductions in mass m and thus increasing the velocity dramatically.

3.3 SECOND-ORDER LINEAR EQUATIONS

Given Newton's Second Law, written as $m\ddot{x}(t) = F(x,t)$, it is clear that second-order ordinary differential equations are ubiquitous in physics. In this section we will focus on linear differential equations. Of course, the form of the force $F(x,t)$ can lead to nonlinear equations and various associated (and fascinating) physical phenomena. More often than not, however, these equations need to be solved numerically.[3]

You should realize that in case where the function is explicitly "missing" in (3.1a), that is

$$\frac{d^2y}{dx^2} = f\left(x, \frac{dy}{dx}\right) \tag{3.7}$$

we can reduce the problem to a first-order equation by writing $v(x) \equiv y'(x)$. In other words, (3.7) becomes

$$\frac{dv}{dx} = f(x,v)$$

which can be solved for $v(x)$ using the techniques in Section 3.2. In fact, this is one of the approaches we used in Section 3.2.4. It is not always an easy matter, though, to integrate the equation for $v(x)$ in order to get $y(x)$.

Now we focus on linear second-order equations. Following (3.1a), we are now working with the functional form

$$f\left(x, y, \frac{dy}{dx}\right) = g(x) - p(x)\frac{dy}{dx} - q(x)y$$

In other words, we are working with differential equations of the form

$$y'' + p(x)y' + q(x)y = g(x)$$

to be solved for the function $y(x)$ with the boundary conditions $y(x_0) = y_0$ and $y'(x_0) = y'_0$. We will use the machinery outlined in Section 3.1.1 and go through some examples. Remember in particular that, for a linear inhomogeneous equation, we can solve the homogeneous equation after finding a particular solution to the inhomogeneous equation, and use this combination to find the coefficients c_i that satisfy the boundary conditions.

[3]Chapter 12 of *Classical Mechanics* by Taylor has a terrific presentation of nonlinear systems and chaos using the driven, damped pendulum.

3.3.1 CONSTANT COEFFICIENTS

Before moving on to physical examples, let's work on what is likely the simplest example of a second-order linear differential equation, namely one that is homogeneous with constant coefficients. That is, an equation of the form

$$ay'' + by' + cy = 0$$

where a, b, and c are real constants. Relying on existence and uniqueness, let's try a solution that looks like an exponential, but with some freedom that we can exploit. We can make the "ansatz" (i.e. "guess") that the solution is of the form $y(x) = e^{\alpha x}$ where α is yet to be determined. Substituting into our differential equation gives

$$\alpha^2 ae^{\alpha t} + \alpha be^{\alpha t} + ce^{\alpha t} = 0$$

Canceling out the factor $e^{\alpha t}$ leaves us with the quadratic equation for α

$$a\alpha^2 + b\alpha + c = 0$$

known as the "characteristic equation." It has the solution

$$\alpha = -\frac{b}{2a} \pm \frac{\sqrt{b^2 - 4ac}}{2a} \equiv \alpha_{1,2}$$

Consequently, we expect that we have two linearly independent solutions

$$y_1(x) = e^{\alpha_1 x} \qquad \text{and} \qquad y_2(x) = e^{\alpha_2 x}$$

If it turns out that $b^2 - 4ac < 0$, then α_1 and α_2 will be complex, and Euler's Formula will give us sines and cosines. Recall Section 2.4.3.

A complication arises if $b^2 - 4ac = 0$. In this case, there is only one linearly independent solution given the $e^{\alpha t}$ ansatz. We have to figure out some other way to get a second solution. Then, of course, if we want to solve an inhomogeneous equation, we need to find a particular solution.

Let's illustrate all this with some specific examples. First, consider the homogeneous equation

$$y'' - y = 0 \qquad \text{with} \qquad y(0) = 2 \qquad \text{and} \qquad y'(0) = -1$$

That is, $a = 1$, $b = 0$, and $c = -1$, giving $\alpha = \pm 1$, in which case $y_1(x) = e^x$ and $y_2(x) = e^{-x}$ so that

$$y(x) = c_1 y_1(x) + c_2 y_2(x) = c_1 e^x + c_2 e^{-x}$$

The boundary conditions tell us that $2 = c_1 + c_2$ and $-1 = c_1 - c_2$ so that $c_1 = 1/2$ and $c_2 = 3/2$. Thus, the complete solution is

$$y(x) = \frac{1}{2}e^x + \frac{3}{2}e^{-x}$$

It is worthwhile to check the Wronskian to see that the two solutions are indeed linearly independent. Following (3.5) we write

$$W(x) = \begin{vmatrix} y_1(x) & y_2(x) \\ y_1'(x) & y_2'(x) \end{vmatrix} = \begin{vmatrix} e^x & e^{-x} \\ e^x & -e^{-x} \end{vmatrix} = -2$$

which of course is nonzero for all x. The two solutions $y_1(x)$ and $y_2(x)$ are linearly independent for all $x \in \mathbb{R}$.

Next consider the similar but inhomogeneous equation

$$y'' - y = 1 \qquad \text{with} \qquad y(0) = 2 \qquad \text{and} \qquad y'(0) = -1$$

with the same boundary conditions. It is simple to guess the particular solution $y_P(x) = -1$. Therefore, the general solution is

$$y(x) = y_P(x) + c_1 y_1(x) + c_2 y_2(x) = -1 + c_1 e^x + c_2 e^{-x}$$

and the boundary conditions tell us that

$$\begin{aligned} 2 &= -1 + c_1 + c_2 \\ \text{and} \quad -1 &= c_1 - c_2 \end{aligned}$$

which are easy enough to solve for $c_1 = 1$ and $c_2 = 2$, giving us the complete solution

$$y(x) = -1 + e^x + 2e^{-x}$$

You should check that this satisfies the differential equation as well as the boundary conditions.

Now let's try a third and final example. Consider the equation

$$y'' + 4y' + 4y = 0$$

for which the characteristic equation becomes $\alpha^2 + 4\alpha + 4 = (\alpha + 2)^2 = 0$. There is only one solution, namely $\alpha = -2$. How can we find the second linearly independent solution so that we can satisfy arbitrary boundary conditions?

Once again, we are rescued thanks to existence and uniqueness. Let's guess that if we modify our one solution by a function $v(x)$, that is $y_2(x) = v(x)e^{-2x}$, and substitute it into our differential equation, then we should be able to find a differential equation for $v(x)$ that we can solve. Doing the work,

$$\begin{aligned} y_2(x) &= v(x)e^{-2x} \\ y_2'(x) &= v'(x)e^{-2x} - 2v(x)e^{-2x} \\ y_2''(x) &= v''(x)e^{-2x} - 2v'(x)e^{-2x} - 2v'(x)e^{-2x} + 4v(x)e^{-2x} \\ &= v''(x)e^{-2x} - 4v'(x)e^{-2x} + 4v(x)e^{-2x} \\ y_2'' + 4y_2' + 4y_2 &= v''(x)e^{-2x} - 4v'(x)e^{-2x} + 4v(x)e^{-2x} \\ &\quad +4v'(x)e^{-2x} - 8v(x)e^{-2x} + v(x)e^{-2x} \\ &= v''(x)e^{-2x} = 0 \end{aligned}$$

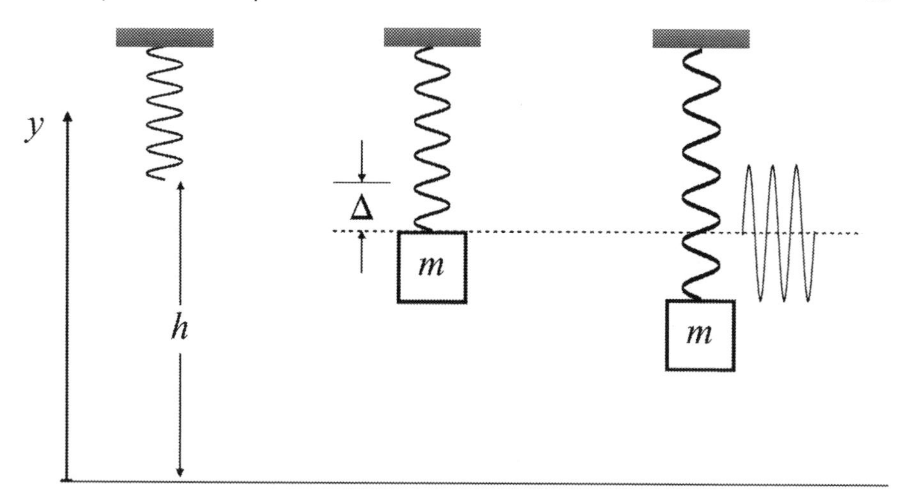

Figure 3.1 A spring with stiffness k hangs from the ceiling. The end of the spring is a height $y = h$ from the floor (left) but a mass m hangs from it, extending it a distance Δ when the mass is not moving (center). The end of the spring is now the "equilibrium point" in height, around which the mass oscillates vertically (right).

Aha! We want a function $v(x)$ that satisfies $v''(x) = 0$. Obviously, that is $v(x) = c_1 + c_2 x$, and the general solution to this homogenous second-order homogeneous equation is

$$y(x) = (c_1 + c_2 x)e^{-2x}$$

We don't have to bother with considering $y_1(x)$ from above, because this approach alone gave us a general solution with two constants. Existence and uniqueness tell us that we are done.

3.4 HARMONIC MOTION IN ONE DIMENSION

Let's now use this machinery to study harmonic motion, a subject we started in Section 2.4.3. It is a very useful example because, in addition to being very important physically, it embodies so much of the mathematics we have discussed to this point, as well as mathematics we have yet to get to.

3.4.1 SIMPLE HARMONIC MOTION

The motion of a mass m subject to a linear restoring force $-kx$, with no other forces in the direction of motion, is called *Simple Harmonic Motion*. This problem was worked out thoroughly in Section 2.4.3, but let's take a moment to couch that discussion in the language of second-order linear differential equations.

Before connecting on to what we saw in Section 2.4.3, however, it is instructive to look at a particular physical situation, namely that of a mass hanging from a spring attached to the ceiling. See Figure 3.1. Using y to measure the vertical (upward) position of the end of the spring, we see that the spring exerts a force $k(h-y)$ on the mass, namely upward (downward) if $y < h$ ($y > h$). There is also the downward weight mg acting on the mass, so Newton's Second Law becomes the differential equation

$$my''(t) = k(h-y) - mg \qquad \text{or} \qquad my''(t) + ky(t) = kh - mg \tag{3.8}$$

This is a second-order inhomogeneous equation, where the homogeneous solution is just that for simple harmonic motion. The particular solution in this case is very simple, since the right side is just a constant. That is, $y_p = h - mg/k$, which itself is simply a constant.

However, there is a neater way to formulate this problem. If y_0 measures the height when the mass is hanging on the spring but at rest, then we define

$$x \equiv -(y - y_0)$$

as the position *away* from the ceiling, measured with respect to this equilibrium point. In this sense, $x(t)$ is just the same as what is shown in Figure 2.4.

In fact, $y_0 = h - \Delta$ where Δ is the amount the string stretches when you hang the mass on it. Since $k\Delta = mg$, we see that (3.8) becomes

$$-mx''(t) + k(h - mg/k - x) = kh - mg \qquad \text{or} \qquad mx''(t) + kx(t) = 0$$

and we recover (2.12). The point is that the weight of the mass has no effect on the oscillatory motion, other than to shift the point about which the oscillations occur.

We solved the differential equation (2.12), which we rewrite as

$$x''(t) + \omega_0^2 x(t) = 0 \qquad x(0) = x_0 \qquad x'(0) = v_0$$

where I have included the initial conditions. This is an example of a second-order linear ODE with constant coefficients, which we discussed in detail in Section 3.3.1. We approached the solution in a couple of ways, one of which was to use the ansatz $x(t) = e^{\alpha t}$. In this case, the characteristic equation is

$$\alpha^2 + \omega_0^2 \qquad \text{so} \qquad \alpha = \pm i\omega_0$$

in which case the general solution is

$$x(t) = c_1 e^{i\omega_0 t} + c_2 e^{-i\omega_0 t}$$

Applying the initial conditions led us to the complete solution (2.16), that is

$$x(t) = R\cos(\omega_0 t + \phi)$$

where (2.15) gives R and ϕ in terms of x_0 and v_0.

Recall also that we used Simple Harmonic Motion as an example of writing a differential equation in terms of scaled variables. See Section 3.1.3.

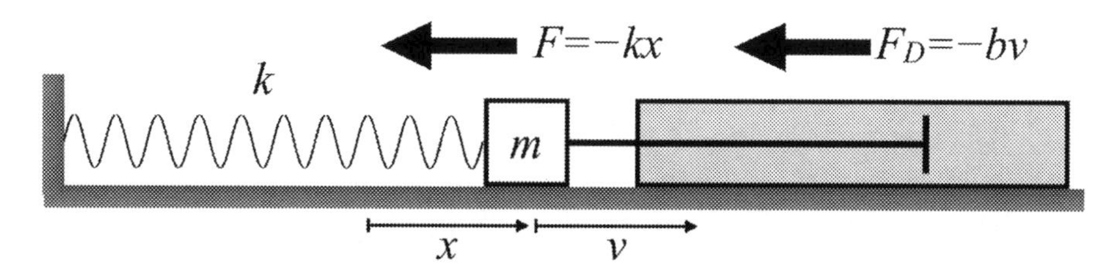

Figure 3.2 The physical situation leading to damped harmonic oscillation. In addition to the restoring force $F = -kx$, there is also a linear damping force $F_D = -bv$ which acts in the direction opposite of the velocity.

3.4.2 DAMPED HARMONIC MOTION

Now let's take the next step and add a (linear) damping force to the mass. See Figure 3.2. Newton's Second Law now says that

$$m\frac{d^2x}{dt^2} = -kx - bv = -kx - b\frac{dx}{dt}$$

If we define $\beta \equiv b/2m$, and keep our definition $\omega_0^2 = k/m$, then the differential equation we need to solve is

$$x''(t) + 2\beta x'(t) + \omega_0^2 x(t) = 0 \tag{3.9}$$

Once again, this is a linear second-order homogeneous differential equation with constant coefficients. Applying our ansatz $x(t) = e^{\alpha t}$ gives the characteristic equation

$$\alpha^2 + 2\beta\alpha + \omega_0^2 = 0 \qquad \text{so} \qquad \alpha = -\beta \pm \sqrt{\beta^2 - \omega_0^2}$$

In the familiar case known as *under damping*, $\beta < \omega_0$ and the argument of the square root is negative. Therefore α is complex and we write $\alpha = -\beta \pm i\omega_1$ where $\omega_1 = \sqrt{\omega_0^2 - \beta^2}$ and the general solution takes the form

$$x(t) = e^{-\beta t}\left(c_1 e^{i\omega_1 t} + c_2 e^{-i\omega_1 t}\right) = Re^{-\beta t}\cos(\omega_1 t + \phi) \tag{3.10}$$

where the constants c_1 and c_2, or R and ϕ, are determined from the initial conditions. Figure 3.3 plots two examples of damped oscillatory motion.

A useful quantity when discussing damped oscillations, especially for very weakly damped oscillations for which $\beta \ll \omega_0$, is the Q (for "Quality") factor, defined as

$$Q = \frac{\omega_0}{2\beta} \tag{3.11}$$

This quantity is best interpreted in terms of the energy of the oscillator. When the mass is at its maximum position, that is the amplitude, all of the oscillator's energy is in potential energy which is proportional to the square of the amplitude. The amplitude in (3.10) decreases like $e^{-\beta t}$, so the energy of the oscillator depends on time as

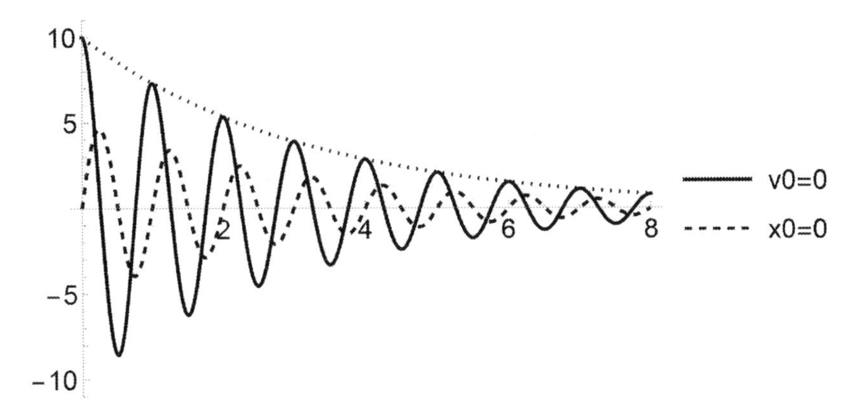

Figure 3.3 Example of under damped harmonic oscillations. The curves are drawn for $\omega_0 = 2\pi$ and $\beta = 0.05\omega_0$. The curve labeled $v_0 = 0$ uses $x_0 = 10$, and the curve labeled $x_0 = 0$ uses $v_0 = 5\omega_0$. The dotted line shows the envelope of the decreasing exponential amplitude, that is $Re^{-\beta t}$ with the notation from the text.

$E(t) = E(0)e^{-2\beta t}$. Therefore, for the weakly damped case, the change in the energy over one period of oscillation is

$$\Delta E = E(t) - E\left(t + \frac{2\pi}{\omega_0}\right) = E(t)\left[1 - e^{-2\beta(2\pi/\omega_0)}\right] \approx E(t)\,2\pi\frac{2\beta}{\omega_0} \qquad \text{for} \qquad \beta \ll \omega_0$$

In other words, for a weakly damped oscillator, the fractional change in energy over one cycle is $\Delta E/E = 2\pi/Q$. A large value of Q means that the oscillator is not very "lossy," and therefore of a higher "quality." We will return to the usefulness of Q when we study driven, damped oscillations in Section 3.4.3.

As discussed in Section 3.3.1, the solution for the *over damped* case when $\beta > \omega_0$ is

$$x(t) = c_1 e^{\alpha_1 t} + c_2 e^{\alpha_2 t} \qquad \alpha_{1,2} = -\beta \pm \sqrt{\beta^2 - \omega_0^2}$$

and the *critically damped* case for $\beta = \omega_0$ has the solution

$$x(t) = (c_1 + c_2 t)e^{-\beta t}$$

Figure 3.4 plots these solutions for the same initial conditions as in Figure 3.3. Of course, the solutions do not oscillate in either of these cases, and the critically damped solution approaches zero most rapidly.

3.4.3 DRIVEN DAMPED HARMONIC MOTION

Now imagine what happens if we "drive" the oscillator by moving the equilibrium point back and forth in an oscillatory manner whose frequency we can control. See Figure 3.5. The other side of the spring from the mass is no longer fixed, but driven

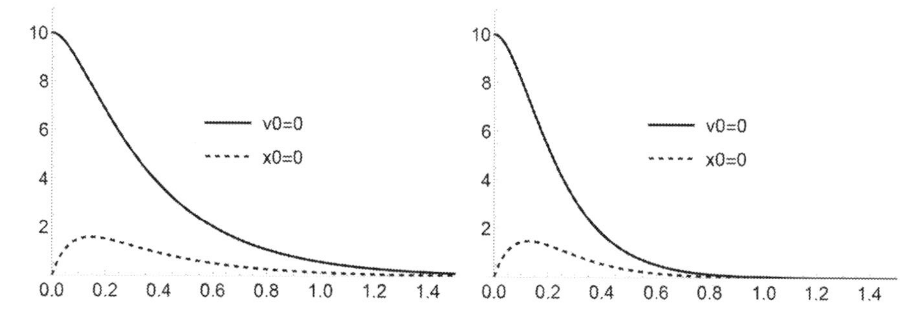

Figure 3.4 Examples of under damped (left) and critically damped (right) motion. The same initial conditions were used as those in Figure 3.3, and once again $\omega_0 = 2\pi$. The under damped case uses the value $\beta = 1.25\omega_0$. Notice that the critically damped case approaches zero more rapidly.

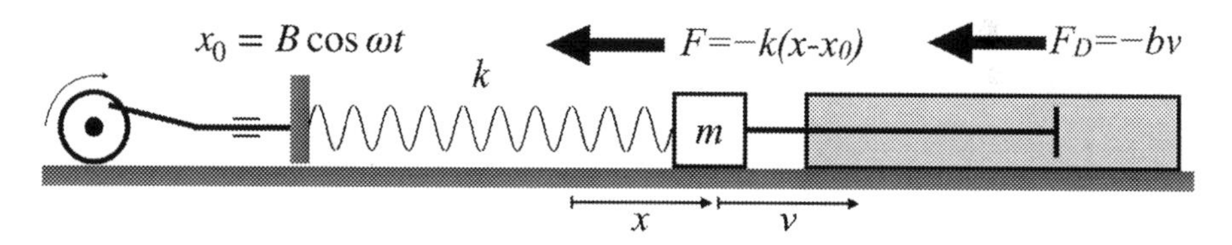

Figure 3.5 The physical situation leading to driven damped harmonic oscillation. The left side attachment point of the spring is no longer fixed, but instead is forced to oscillate sinusoidally by driving it to move with angular frequency ω. This modifies the restoring force of the spring on the mass to be $F = -k(x - x_0)$ where $x_0 = B\cos\omega t$.

along a point $x_0 = B\cos\omega t$. This means that the force from the spring on the mass is $F = -k(x - x_0)$ and the differential equation that describes the system is given by

$$mx''(t) + bx'(t) + kx(t) = kx_0 = kB\cos\omega t$$

Defining $\beta = b/2m$ and $\omega_0^2 = k/m$ as before, and also defining $\gamma = kB/m$, we have

$$x''(t) + 2\beta x'(t) + \omega_0^2 x(t) = \gamma\cos\omega t \tag{3.12}$$

This is a linear, second order, inhomogeneous differential equation with constant coefficients, a case that we also treated in Section 3.3.1.

The only tricky part to solving this problem is to find a particular solution. Having that, we just need to add it to the general solution for the homogeneous equation, and then find the two constants by applying the initial conditions. Looking at (3.12), however, it is by no means obvious what is the particular solution. On the other hand, it is probably some linear combination of $\cos\omega t$ and $\sin\omega t$ so we write

$$x_P(t) = a\cos\omega t + b\sin\omega t$$

and insist that this form solve (3.12) in order to determine a and b.[4] We find

$$\begin{aligned}
&-a\omega^2 \cos \omega t - b\omega^2 \sin \omega t \\
&-2\beta a\omega \sin \omega t + 2\beta b\omega \cos \omega t \\
&+\omega_0^2 a \cos \omega t + \omega_0^2 b \sin \omega t = \gamma \cos \omega t
\end{aligned}$$

Gathering up terms proportional to $\cos \omega t$ and $\sin \omega t$ give us the equations

$$\begin{aligned}
(\omega_0^2 - \omega^2)a + 2\beta\omega b &= \gamma \\
-2\beta\omega a + (\omega_0^2 - \omega^2)b &= 0
\end{aligned}$$

Solving this pair of equations for a and b is straightforward, albeit tedious. You find

$$a = \frac{\gamma\left(\omega_0^2 - \omega^2\right)}{4\beta^2\omega^2 + \left(\omega_0^2 - \omega^2\right)^2} \qquad \text{and} \qquad b = \frac{2\beta\gamma\omega}{4\beta^2\omega^2 + \left(\omega_0^2 - \omega^2\right)^2}$$

It makes sense to write

$$A(\omega) = \sqrt{a^2 + b^2} = \frac{\gamma}{\sqrt{4\beta^2\omega^2 + \left(\omega_0^2 - \omega^2\right)^2}} \tag{3.13a}$$

$$\text{and} \qquad \tan\Phi(\omega) = \frac{b}{a} = \frac{2\beta\omega}{\omega_0^2 - \omega^2} \tag{3.13b}$$

in which case we have

$$x_P(t) = A\cos\Phi\cos\omega t + A\sin\Phi\sin\omega t = A\cos(\omega t - \Phi)$$

Now we can use the results of Section 3.4.2 to write the complete solution to (3.12) as

$$x(t) = c_1 e^{\alpha_1 t} + c_2 e^{\alpha_2 t} + A\cos(\omega t - \Phi)$$

where $\alpha_1 = -\beta + \sqrt{\beta^2 - \omega_0^2}$ and $\alpha_2 = -\beta - \sqrt{\beta^2 - \omega_0^2}$ are complex for the oscillating (under damped) case ($\beta < \omega_0$) or real for the over damped case ($\beta > \omega_0$), and

$$x(t) = (c_1 + c_2 t)e^{-\beta t} + A\cos(\omega t - \Phi)$$

for the critically damped case ($\beta = \omega_0$). For all cases, we find c_1 and c_2 by applying initial conditions $x(0) = x_0$ and $x'(0) = v_0$.

Figure 3.6 shows the motion of an oscillator that starts from rest at $x = 0$, but is set in motion by the driving force. The motion is plotted for three different driving frequencies. The amplitude is much larger for $\omega = \omega_0$ than for the other two choices. This is not unexpected, given the denominator in our expression for $A(\omega)$.

The figure also includes an expanded view of the motions for $\omega = 0.5\omega_0$ and $\omega = 1.5\omega_0$. Notice how the motion appears somewhat irregular for the first several

[4]A much slicker way to do this is to let $x(t)$ be complex, and write the right side of the equation as $\gamma e^{i\omega t}$. I'm going to take the more straightforward approach here.

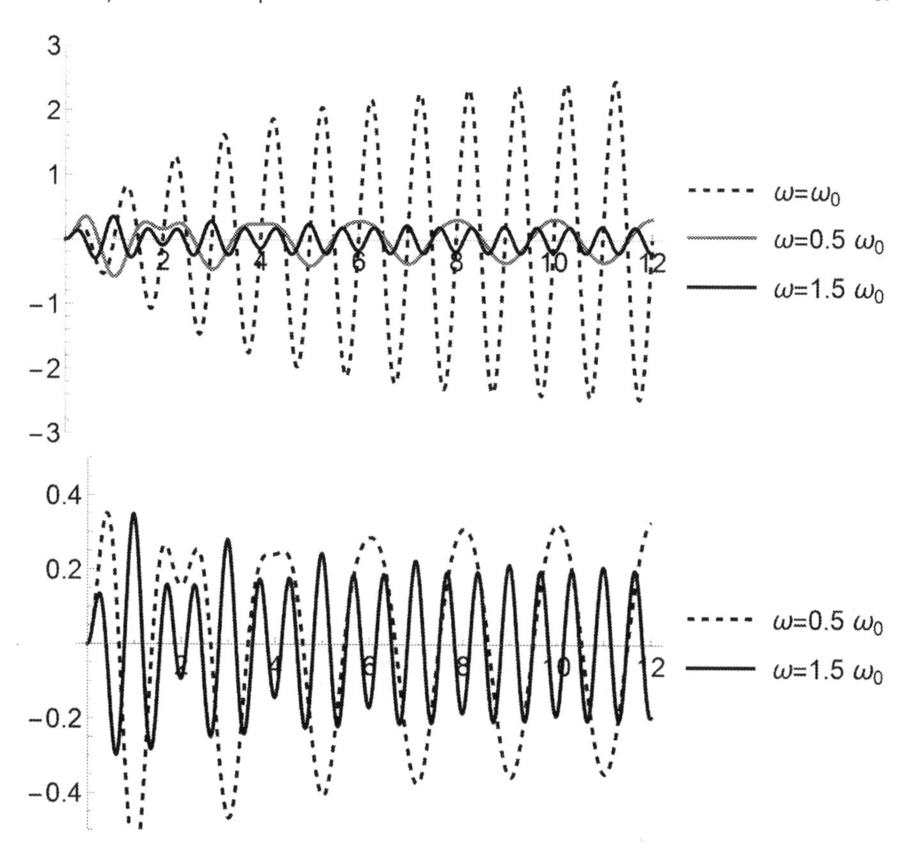

Figure 3.6 Plot of the motion $x(t)$ of an underdamped oscillator driven at three different frequencies ω relative to the natural frequency ω_0. The plot assumes $\omega_0 = 2\pi$, $\beta = 0.05\omega_0$ and uses initial conditions $x(0) = 0$ and $x'(0) = 0$. The driving amplitude is $\gamma = 10$. The bottom plot is an expanded view of the "off resonance" $\omega = 0.5\omega_0$ and $\omega = 1.5\omega_0$ curves on the top, and clearly illustrate the behavior of the transient parts of the solutions.

periods before settling into a steady motion at the driving frequency. This is the effect of the *transients* that are the elements of the homogeneous solution. The transients die away with a time constant $1/\beta$.

The large amplitude in Figure 3.6 when $\omega \approx \omega_0$ is the familiar phenomenon known as *resonance*. When you drive any underdamped oscillator at a frequency close to its natural frequency, it responds with a large amplitude. Figure 3.7 shows explicitly how the amplitude depends on the driving frequency, showing a "resonance peak." Note also that the relative phase between the response and the driving signal varies with frequency as well. For low frequencies, the mass responds directly in phase with the driver, while for frequencies rather larger than the natural frequency, the response is 180° out of phase.

Resonance phenomena are ubiquitous in nature. The mathematics of an electrical oscillator made from a resistor, capacitor, and inductor obeys exactly the same

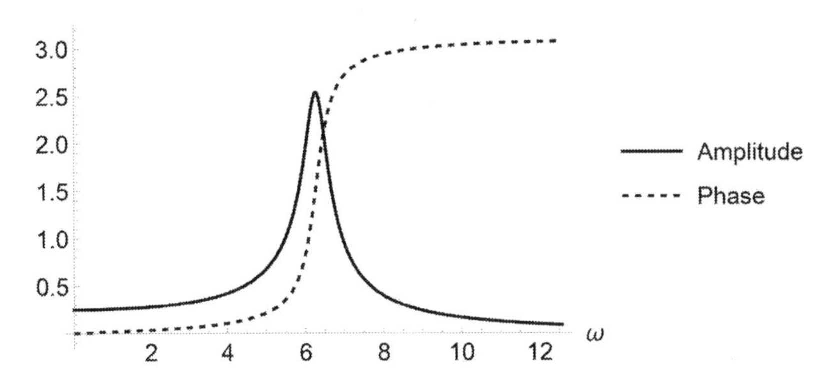

Figure 3.7 The amplitude $A(\omega)$ and phase $\Phi(\omega)$ for the particular solution $x_P(t)$, using the same parameters as in Figure 3.6. The peak in the amplitude is for ω near ω_0, which is where the phase Φ rises through 90°.

differential equations that we have studied here. Quantum mechanical scattering is another well-known system that displays resonance under the right conditions. There are also many examples in the response of living systems to oscillatory driving functions. In all cases, there is a peak in the amplitude, and a relative phase between the drive and response which rises from zero through 90° at resonance, and levels out to 180° out of phase at high frequencies.

Finally, consider the amplitude function $A(\omega)$ from (3.13a) for the weak damping situation $\beta \ll \omega_0$. The peak in the amplitude is near $\omega = \omega_0$ as see in Figure 3.7. We can also determine the "width" of the peak by calculating the values of ω for $A(\omega) = A_{\text{max}}/\sqrt{2} \approx \gamma/2\sqrt{2}\beta\omega_0$. (We choose a fall by $\sqrt{2}$ because it is when the energy falls to one half its maximum value.) This occurs when

$$\sqrt{4\beta^2\omega^2 + (\omega_0^2 - \omega^2)^2} = 2\sqrt{2}\beta\omega_0$$

Substituting $\omega \approx \omega_0$ in the first term under the square root, we find $(\omega_0^2 - \omega^2)^2 = 4\beta^2\omega_0^2$. Solving for ω gives

$$\omega = \left[\omega_0^2 \pm 2\beta\omega_0\right]^{1/2} = \omega_0\left[1 \pm \frac{2\beta}{\omega_0}\right]^{1/2} \approx \omega_0\left[1 \pm \frac{\beta}{\omega_0}\right]$$

Therefore the fractional "width" of the resonance $\Delta\omega/\omega_0 = 2\beta/\omega_0 = 1/Q$, the "quality factor" from (3.11). This is a useful quantity for describing the performance of many resonant systems.

3.5 SERIES SOLUTIONS FOR SECOND-ORDER EQUATIONS

So far, even though we've been talking about differential equations in general terms, the only solution we've come up with for linear second-order ODE's is for the case where the homogeneous equation has constant coefficients. Indeed, there are many other examples of important second-order linear ODE's in the physical sciences. Are there general approaches we can use to address them?

We learned in Chapter 2 that we can represent functions by infinite power series. We can take an approach, then, where the differential equation defines the function, and then we can use a power series to try and craft the solution. In fact, this approach works very well.

The basic idea is the following. Start with the homogeneous equation

$$r(x)y''(x) + p(x)y'(x) + q(x)y(x) = 0$$

and write the solution as a power series with unknown coefficients, that is

$$y(x) = a_0 + a_1 x + a_2 x^2 + \cdots = \sum_{n=0}^{\infty} a_n x^n$$

Substitute this form into the differential equation and manipulate the terms to find a "recursion relation" that relates a coefficient a_{n+2} to coefficients a_{n+1} and/or a_n. The coefficients a_0 and a_1 remain as the constants of integration, which get determined using the boundary conditions $y(0) = y_0$ and $y'(0) = y'_0$.

This approach works for a lot of problems, and we'll be discussing them soon. It doesn't always work, though. Remember that you have existence and uniqueness on your side, so variations of this approach can be used if necessary.

One variation is that the boundary values might be specified for $x \to \pm\infty$, a common situation in Quantum Mechanics. This variation tends to lead to the series being truncated, so the solution is actually a polynomial.

Another variation, known as the *Method of Frobenius*, is to write

$$y(x) = x^s \sum_{n=0}^{\infty} a_n x^n = \sum_{n=0}^{\infty} a_n x^{s+n}$$

and use the differential equation to constrain, or determine, the value of s.

3.5.1 SIMPLE EXAMPLES

Let's do some examples where we know what the answer has to be, and use these to illustrate how the series solution approach works. Nothing about the approach is restricted to second-order equations, so let's start with

$$y'(x) = y(x)$$

Of course, the solution is $y(x) = ce^x$ where c is a constant. Inserting the series we have

$$\sum_{n=0}^{\infty} n a_n x^{n-1} = \sum_{n=0}^{\infty} a_n x^n \tag{3.14}$$

Now look at the sum on the left. The first term ($n = 0$) is manifestly zero, so we can start the sum from $n = 1$ and then switch to a dummy index $m = n - 1$. This gives

$$\sum_{n=0}^{\infty} na_n x^{n-1} = \sum_{n=1}^{\infty} na_n x^{n-1} = \sum_{m=0}^{\infty} (m+1)a_{m+1}x^m$$

Switching the dummy index back to n and rearranging the terms in (3.14) we get

$$\sum_{n=0}^{\infty} [(n+1)a_{n+1} - a_n]x^n = 0 \tag{3.15}$$

Sometimes this "dummy index" switching can be confusing, so let's present it a little differently. Writing

$$\begin{aligned} y'(x) &= \sum_{n=0}^{\infty} na_n x^{n-1} = (0)a_0 x^{-1} + (1)a_1 x^0 + (2)a_2 x^1 + (3)a_3 x^2 + \cdots \\ &= \sum_{n=0}^{\infty} (n+1)a_{n+1}x^n \end{aligned}$$

it should be clear that we can write the sum as (still) starting at $n = 0$ but with powers x^n. This is only possible, however, because the first term in the original sum is zero.

Now, in order for (3.15) to be valid for all values of x, we must require

$$a_{n+1} = \frac{1}{n+1}a_n$$

This is the "recursion relation" mentioned above. It tells us how to get any coefficient in terms of the one before. If we set $a_0 = c$ then it is clear that

$$a_1 = \frac{1}{1}c \qquad a_2 = \frac{1}{2}a_1 = \frac{1}{2 \cdot 1}c \qquad a_3 = \frac{1}{3}a_2 = \frac{1}{3 \cdot 2 \cdot 1}c$$

and so forth. In other words, $a_n = c/n!$ for all n. The resulting solution is

$$y(x) = \sum_{n=0}^{\infty} \frac{1}{n!}cx^n = c\sum_{n=0}^{\infty} \frac{1}{n!}x^n = ce^x$$

and indeed we get the solution we expected.

We'll do one more simple example to see how this works for a second-order equation. Let's use the series approach to find the solution to

$$y''(x) = -y(x)$$

which we know to be $y(x) = c_1 \cos x + c_2 \sin x$. Substituting the series form and doing the same manipulations as before ends up with

$$\sum_{n=0}^{\infty} [(n+2)(n+1)a_{n+2} + a_n]x^n = 0 \qquad \text{so} \qquad a_{n+2} = -\frac{1}{(n+2)(n+1)}a_n$$

This time, the recursion relation skips over the prior term, which means that we are free to specify both a_0 and a_1. Writing $a_0 = c_1$ and $a_1 = c_2$ we get

$$\begin{aligned}
a_2 &= -\frac{1}{2\cdot 1}c_1 \qquad \text{and} \qquad a_4 = -\frac{1}{4\cdot 3}a_2 = +\frac{1}{4!}c_1 \quad \cdots \\
\text{so} \qquad a_{2n} &= (-1)^n\frac{1}{n!} \quad \text{for } n = 0,1,2,3,\ldots \\
\text{and also} \qquad a_3 &= -\frac{1}{3\cdot 2}c_2 \qquad \text{and} \qquad a_5 = -\frac{1}{4\cdot 4}a_3 = +\frac{1}{5!}c_2 \quad \cdots \\
\text{so} \qquad a_{2n+1} &= (-1)^n\frac{1}{n!} \quad \text{for } n = 0,1,2,3,\ldots
\end{aligned}$$

The solution naturally separates into two series, namely

$$\begin{aligned}
y(x) &= c_1\left[1 - \frac{1}{2!}x^2 + \frac{1}{4!}x^4 + \cdots\right] + c_2\left[x - \frac{1}{3!}x^3 + \frac{1}{5!}x^5 + \cdots\right] \\
&= c_1 \cos x + c_2 \sin x
\end{aligned}$$

and once again we get the correct solution.

Many physical problems result in second-order linear ODE's whose solutions can only be written in terms of series. Many of these solutions are given special names. A few of the most important examples are discussed in Section 3.6.

3.5.2 THE QUANTUM MECHANICAL HARMONIC OSCILLATOR

Quantum Mechanics can be formulated in terms of differential equations. In particular, the Schrödinger Equation is an approach to Quantum Mechanics where one solves for the *wave function* $\psi(x)$ for a particle of mass m acted on by a potential energy function $V(x)$ by solving the differential equation

$$-\frac{\hbar^2}{2m}\frac{d^2\psi}{dx^2} + V(x)\psi(x) = E\psi(x)$$

subject to some boundary conditions. For solutions that are localized in some region of x, so-called "bound state solutions," it turns out that solutions are only possible for discrete values of the total energy E. That is, the energies are "quantized," hence the name of this field. The boundary conditions do not typically make it possible to find all of the integration parameters, but an additional *normalization* constraint is included, namely that the integral of $|\psi(x)|^2$ over all x must equal unity. This allows $|\psi(x)|^2$ to be interpreted as a "probability density."

Let's take on the solution for the simple harmonic oscillator, where $V(x) = kx^2/2 = m\omega_0^2x^2/2$ is the potential energy function. (We know from our past work that $\omega_0^2 = k/m$ is a useful parameter, so let's put it in at the start.) The differential equation is

$$-\frac{\hbar^2}{2m}\frac{d^2\psi}{dx^2} + \frac{1}{2}m\omega_0^2x^2\psi(x) = E\psi(x)$$

The function $\psi(x)$ is subject to the boundary condition that $\psi(x \to \pm\infty) = 0$.

First, it is a good idea to change variables from x to a new variable y equal to x divided by some length scale. Since finding the energy will be part of the problem, let's not use E to find a length scale, and instead form it out of $\hbar$, m, and ω_0. Writing the length scale as $\hbar^x m^y \omega_0^z$, we need

$$L = [\hbar]^x \cdot [m]^y \cdot [\omega_0]^z = L^{2x} M^x T^{-x} \cdot M^y \cdot T^{-z} = L^{2x} M^{x+y} T^{-x-z}$$

so $x = 1/2$ and $y = -1/2 = z$. Therefore we write $y = x/\sqrt{\hbar/m\omega_0}$ and

$$-\frac{\hbar^2}{2m}\frac{m\omega_0}{\hbar}\frac{d^2\psi}{dy^2} + \frac{1}{2}m\omega_0^2\frac{\hbar}{m\omega_0}y^2\psi(y) = E\psi(y)$$

Multiplying through by $2/\hbar\omega_0$ and defining $\varepsilon \equiv 2E/\hbar\omega_0$, and doing a little rearranging, we arrive at the second-order linear homogeneous differential equation

$$\psi''(y) + (\varepsilon - y^2)\psi(y) = 0 \tag{3.16}$$

This equation looks innocuous enough, but in fact its solution will take some gymnastics, even before we get to applying the series solution. Solving this will be a good illustration of how physicists make use of "flying by the seat of the pants" to come up with solutions to differential equations.

The fact that we need $\psi(y)$ to go to zero for large $\pm y$ suggests a good first step. Since ε is a constant, we might first consider the differential equation

$$\psi''(y) - y^2\psi(y) = 0$$

to understand the dependence on y for $y \to \pm\infty$. (We typically call this the "asymptotic dependence.") This solution to this equation is also not available in an analytic form, but if we consider the function

$$\psi(y) = e^{-y^2/2} \qquad \text{then} \qquad \psi''(y) - y^2\psi(y) = -e^{-y^2/2} \to 0 \qquad \text{as} \qquad y \to \pm\infty$$

This suggests the "asymptotic behavior" of $\psi(y)$ might behave something like $e^{-y^2/2}$. We can remove this behavior by writing

$$\psi(y) = e^{-y^2/2}h(y) \tag{3.17}$$

and then inserting this into (3.16) to find the differential equation satisfied by $h(y)$.

I admit this all sounds cockamamie. However, don't forget that existence and uniqueness are there for us, so any way that works is fine.

So let's use (3.17) in (3.16) and see what happens. The derivatives are

$$\begin{aligned}
\psi'(y) &= e^{-y^2/2}h'(y) - ye^{-y^2/2}h(y) \\
\psi''(y) &= e^{-y^2/2}h''(y) - 2ye^{-y^2/2}h'(y) + y^2e^{-y^2/2}h(y) - e^{-y^2/2}h(y)
\end{aligned}$$

Inserting this into (3.16) gives

$$e^{-y^2/2}h''(y) - 2ye^{-y^2/2}h'(y) - e^{-y^2/2}h(y) + \varepsilon e^{-y^2/2}h(y) = 0$$

Dividing out the exponential factor leaves us with

$$h''(y) - 2yh'(y) + \lambda h(y) = 0 \qquad \text{where} \qquad \lambda \equiv \varepsilon - 1 \tag{3.18}$$

Mathematicians have given this second-order linear ODE a name, the Hermite Equation. We can use the series approach to solve this equation. Proceeding as usual we have

$$\begin{aligned} h(y) &= \sum_{n=0}^{\infty} a_n y^n \\ h'(y) &= \sum_{n=0}^{\infty} n a_n y^{n-1} \qquad \text{so} \qquad yh'(y) = \sum_{n=0}^{\infty} n a_n y^n \\ h''(y) &= \sum_{n=0}^{\infty} n(n-1) a_n y^{n-2} = \sum_{n=0}^{\infty} (n+2)(n+1) a_{n+2} y^n \end{aligned}$$

and the Hermite Equation (3.18) becomes

$$\sum_{n=0}^{\infty} \left[(n+2)(n+1)a_{n+2} - (2n-\lambda)a_n\right] y^n = 0$$

which gives the recursion relation

$$a_{n+2} = \frac{2n-\lambda}{(n+2)(n+1)} a_n \tag{3.19}$$

and we get two independent series, one with even powers of y, and one with odd powers. (There is actually some interesting Quantum Mechanics that has to do with this observation, concerning parity symmetry, but I will leave that to your Quantum Mechanics course.)

Now this solution would actually present a difficulty. Remember that we need $\psi(x) \to 0$ for $x \to \pm\infty$. It would seem that (3.17) covers this, provided that $h(y)$ does not grow too quickly. However, consider (3.19) for large values of y, where the series is dominated by $n \gg 1$. In this case $a_{n+2} \to (2n/n^2)a_n = (2/n)a_n$. For the series with even values of $n = 2m$, the series becomes something proportional to

$$\sum_{m \gg 1} \frac{1}{m!} y^{2m} = e^{y^2}$$

From (3.17), this implies that $\psi(y) \to e^{y^2/2}$ as $y \to \pm\infty$, and there is no way that we can meet the boundary condition. (We can make the same argument for the series with odd n, just by factoring out y.)

Nevertheless, there is a solution to this problem. If the recursion relation says that the series *terminates* at some value of n, then $h(y)$ is not an infinite series, but rather a polynomial of degree n. This happens if, for some n, the numerator of (3.19) vanishes. That is

$$2n - \lambda = 2n - (\varepsilon - 1) = 0$$

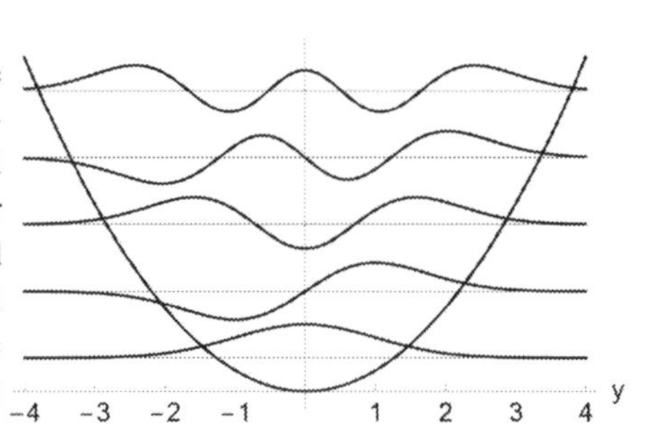

Figure 3.8 The wave functions $\psi_n(y)$ for the quantum mechanical simple harmonic oscillator, superimposed on the potential energy and energy levels. Wave functions are shown for $n = 0, 1, 2, 3, 4$ and are plotted vertically shifted so that $\psi = 0$ corresponds to the energy level $E_n = (n + 1/2)\hbar\omega_0$. Notice that $\psi_n(-y) = \psi_n(y)$ for n even, and that $\psi_n(-y) = -\psi_n(y)$ for n odd.

Since $\varepsilon \equiv 2E/\hbar\omega_0$, this tells us that the allowed energies of the Quantum Mechanical Simple Harmonic Oscillator are

$$E = \frac{\hbar\omega_0}{2}(2n+1) = \left(n + \frac{1}{2}\right)\hbar\omega_0 \equiv E_n$$

The energies of the oscillator are "quantized" into evenly spaced values of $\hbar\omega_0$. This is a profound and familiar result in Quantum Mechanics, which you will see again in physics.

For any value of n, the finite series $h_n(x)$ is called the Hermite Polynomial of order n. The wave functions $\psi(y)$ can then be written as

$$\psi(y) = N_n h_n(y) e^{-y^2/2} \equiv \psi_n(y)$$

where N_n is a "normalization constant" that is set by a different assumption of Quantum Mechanics, namely that the integral of $|\psi(y)|^2$ over all y is unity. There are ways to determine a formula for N_n, but I won't cover that here. The first several Hermite polynomials are

$$\begin{aligned} h_0(y) &= 1 \\ h_1(y) &= 2y \\ h_2(y) &= 4y^2 - 2 \\ h_3(y) &= 8y^3 - 12y \\ h_4(y) &= 16y^4 - 48y^2 + 12 \end{aligned}$$

Figure 3.8 shows the first five wave functions plotted on top of the energy levels. The "parity" $(-1)^n$ of the wave functions is apparent, as is the fact that $\psi_n(y) \to 0$ for $y \to \pm\infty$. In fact, the inflection point where the wave function turns to decreasing from oscillatory always occurs at the "classical turning point" where the total energy equals the potential energy.

We have only scratched the surface of this very important problem in Quantum Mechanics. You will learn more in other courses, but note the lesson here that sometimes the infinite series solution will be truncated to a polynomial. We will encounter an example of this in the next section when we discuss the Legendre Equation.

3.6 SOME IMPORTANT SPECIAL FUNCTIONS

There are many physical situations that give rise to specific linear second-order ordinary differential equations. The important ones all have names, usually based on the mathematician who popularized them and their solutions. Oftentimes, the solutions to these equations can only be expressed as infinite series, or series truncated to become polynomials. The solutions generally go by the same name as the differential equations they solve. We saw one example of this in Section 3.5.2 with the Hermite Equation and Hermite Polynomials.

Section 4.2.4 will discuss Laplacian operator $\vec{\nabla}^2$, which leads to partial differential equations in different areas of physics. For example, $\vec{\nabla}^2 V = 0$, called Laplace's Equation, is used to derive the electric potential in the presence of static charges, along with some boundary conditions. In Quantum Mechanics, $\vec{\nabla}^2 \psi = -k^2 \psi$ is solved to find the wave function for a free particle with momentum $\hbar k$. These are important examples because, as we will learn later in Section 4.5.1, a technique called "Separation of Variables" leads to second-order ordinary differential equations in the spatial coordinates.

We call the solutions to these specific linear second-order ODE's "Special Functions" and these are the subject of this section. In many cases, these functions cannot be written in closed form but instead are written as infinite series.

Of course, there is nothing "special" about these Special Functions. They are just defined in terms of the differential equations they solve. We could just as well have defined e^x as the solution to $y'(x) = y$, and $\cos x$ and $\sin x$ as the linearly independent solutions of $y''(x) = -y$, but we didn't. In this section, we will focus on the Bessel Functions of integer order $m \geq 0$ $J_m(x)$, the Spherical Bessel Functions $j_\ell(x)$ and $n_\ell(x)$, and the Legendre Polynomials $P_\ell(x)$. It turns out that $j_\ell(x)$ and $n_\ell(x)$ can be written in closed form using $\cos x$ and $\sin x$, and the Legendre Polynomials are just that, polynomials, but the $J_m(x)$ can only be written as an infinite series.

Before we get into specific special functions, however, we need to do a little more work on the general theory of second-order linear ODE's.

3.6.1 ORDINARY AND SINGULAR POINTS

We return to writing our general second-order homogeneous linear ODE as

$$r(x)y''(x) + p(x)y'(x) + q(x)y(x) = 0$$

In principle, we can just write this equation as

$$y''(x) + \frac{p(x)}{r(x)}y'(x) + \frac{q(x)}{r(x)}y(x) = 0$$

but we have to be careful. For some value of $x = x_0$ at which $r(x_0) = 0$, we cannot just assume that the second equation has a solution that is the same as for the first. Such points $x = x_0$ are called *singular points*, whereas other values of x are called *ordinary points*. If

$$\lim_{x \to x_0} (x - x_0) \frac{p(x)}{r(x)} \quad \text{and} \quad \lim_{x \to x_0} (x - x_0)^2 \frac{q(x)}{r(x)}$$

are finite, then $x = x_0$ is called a *regular singular point*. For regular singular points, the functions $(x - x_0)p(x)/r(x)$ and $(x - x_0)^2 a(x)/r(x)$ will have well behaved Taylor expansions about $x = x_0$, so a series solution approach can be pursued.

We will only be discussing solutions of this class for regular singular points. Solutions for differential equations about irregular singular points is an advanced topic that I will leave for a mathematics course.

Let's examine two of the equations we will study in this section with respect to their singular points. First consider Bessel's Equation, namely

$$x^2 y''(x) + xy'(x) + (x^2 - \nu^2)y(x) = 0 \tag{3.20}$$

Clearly, $x = 0$ is a singular point. However, since

$$\lim_{x \to 0} x \frac{x}{x^2} = \lim_{x \to 0} 1 = 1 \quad \text{and} \quad \lim_{x \to 0} x^2 \frac{x^2 - \nu^2}{x^2} = -\nu^2$$

are both finite, $x = 0$ is a regular singular point and we can go ahead and try to build a series solution for Bessel's Equation. We will do this in Section 3.6.2.

Now also consider the Legendre Equation, that is

$$(1 - x^2)y''(x) - 2xy'(x) + \alpha(\alpha + 1)y(x) = 0 \tag{3.21}$$

In this case there are two singular points, at $x = \pm 1$. However, since

$$\begin{aligned} \lim_{x \to 1} (x-1) \frac{-2x}{1-x^2} = \lim_{x \to 1} (x-1) \frac{-2x}{(1-x)(1+x)} &= 2 \\ \text{and} \quad \lim_{x \to 1} (x-1)^2 \frac{\alpha(\alpha+1)x}{(1-x)(1+x)} &= 0 \end{aligned}$$

are both finite, $x = +1$ is a regular singular point. Similarly for $x = -1$.

Of course, if $x = x_0$ is a singular point, we can always build a series solution around a different value of x, in which case we don't need to be concerned whether the point is regular or irregular. The behavior of the solution at $x = x_0$, though, is likely to be peculiar.

3.6.1.1 Euler equations

There is an instructive class of second-order ODE's with regular singular points called Euler Equations (not to be confused with Euler's Formula from Section 2.4). Euler Equations take the form

$$x^2 y''(x) + \alpha xy'(x) + \beta^2 y(x) = 0 \tag{3.22}$$

It should be clear that there is a regular singular point at $x = 0$. (We could write a slightly different equation with $(x - x_0)$ instead of x in front of the first and second terms, and get a regular singular point at $x = x_0$.)

It should also be clear that a series solution to this equation would not work. No recursion relation comes from inserting the series into the equation, because the powers of x in front of $y''(x)$ and $y'(x)$ bring all powers of x back up to x^n.

Nevertheless, this equation has a straightforward solution that brings to mind the ansatz we used for the second-order ODE with linear coefficients. Inserting $y(x) = x^r$, where r is a constant to be determined, we get the quadratic equation

$$r(r-1) + \alpha r + \beta = r^2 + (\alpha - 1)r + \beta^2 = 0$$

which is to be solved for r. Evidently, the two solutions for r are

$$r = \frac{-(\alpha - 1) \pm \sqrt{(\alpha - 1)^2 - 4\beta^2}}{2}$$

There are three obvious cases. For $(\alpha - 1)^2 > 4\beta^2$, there are two real values $r = r_1$ and $r = r_2$ for which x^r is a solution. That is, the general solution is

$$y(x) = c_1 x^{r_1} + c_2 x^{r_2}$$

If $(\alpha - 1)^2 < 4\beta^2$, then the roots are complex, and we make use of $x^{\mu + i\eta} = e^{(\mu + i\eta)\log x}$ to write the solutions. If the two values of r are equal, we need to figure out some way to get a second equation, but I'll leave this for a homework problem.

The lesson here is that for a regular singular point, including a factor of x^r in the solution, where r is to be determined, may be a useful approach to be included in the series expansion. This approach is sometimes called the Method of Frobenius.

3.6.2 BESSEL FUNCTIONS

A differential equation ubiquitous to physics problems is Bessel's Equation (3.20), which we reproduce here:

$$x^2 y''(x) + xy'(x) + (x^2 - \nu^2)y(x) = 0 \tag{3.23}$$

where ν is a constant, known as the *order*. As discussed above, $x = 0$ is a regular singular point, so we should pursue a series solution of the form

$$y(x) = x^r \sum_{n=0}^{\infty} a_n x^n = \sum_{n=0}^{\infty} a_n x^{r+n} \tag{3.24}$$

where we expect (3.23) to constrain r as well as determine the coefficients a_n. Inserting this form into Bessel's Equation gives us

$$\sum_{n=0}^{\infty} \left\{ \left[(r+n)(r+n-1) + (r+n) - \nu^2 \right] a_n x^{r+n} + a_n x^{r+n+2} \right\} = 0 \tag{3.25}$$

Notice that we would be in trouble if we didn't include the factor x^r. Setting $r=0$ gives

$$\sum_{n=0}^{\infty} \left\{(n^2-\nu^2)a_n x^n + a_n x^{n+2}\right\} = 0$$

which implies that $a_0 = 0$ and $a_1 = 0$, based on the x^0 and x^1 terms. Deriving a recursion relation gives a_{n+2} in terms of a_n, so all the a_n end up as zero. This would all be nonsense!

We can simplify (3.25) to get

$$\sum_{n=0}^{\infty} \left\{\left[(r+n)^2-\nu^2\right] a_n x^{r+n} + a_n x^{r+n+2}\right\} = 0 \tag{3.26}$$

In this case, for the $n=0$ term, we have $r^2-\nu^2=0$ which gives us two possibilities for r, namely $r=\pm\nu$. Setting the coefficient of the $n=1$ term to zero gives

$$[(r+1)^2-\nu^2]a_1 = [\pm 2\nu+1]a_1 = 0$$

which implies that $a_1=0$, unless $\nu=\pm 1/2$, a case we will deal with in Section 3.6.3.

It should be clear that we indeed get two solutions for $y(x)$, one each for $r=\pm\nu$. Following the recursion relation we will derive from (3.26) results in the two functions we call $J_{\pm\nu}(x)$, known as *Bessel Functions of the first kind.* The theory associated with Bessel Functions is extensive, but we will not go very far into it. Nearly all cases in physics that involve Bessel Functions result in functions where ν is either a positive integer, or a positive half-integer.

At this point, we will specialize to the case where $\nu = m \in \mathbb{Z}$, an integer. I will tell you at the outset, but not prove, that for this case, the solutions $J_m(x)$ and $J_{-m}(x)$ are *not* linearly independent. A different solution can be identified, though, usually written as $Y_m(x)$ and called the *Bessel Function of the Second Kind*, which is independent. I am leaving this very interesting mathematics for some later course you will hopefully take.

With the first two terms of the sum being zero, we can rewrite the first term in curly brackets in (3.26) as

$$x^m \left\{\sum_{n=2}^{\infty} \left[(m+n)^2-m^2\right] a_n x^n\right\} = x^m \left\{\sum_{p=0}^{\infty} \left[(m+p+2)^2-m^2\right] a_{p+2} x^{p+2}\right\}$$

where we switch to the dummy index $p=n-2$. Switching p back n, we replace the above expression in (3.26) and divide out x^m to get

$$\sum_{n=0}^{\infty} \left\{\left[(m+n+2)^2-m^2\right] a_{n+2} + a_n\right\} x^{n+2} = 0$$

giving us the recursion relation

$$a_{n+2} = -\frac{1}{(m+n+2)^2-m^2}\, a_n = -\frac{1}{(n+2)(2m+n+2)}\, a_n$$

Since $a_1 = 0$, the sum is only over even n. Therefore we define $n = 2k$ and sum over all integers $k \geq 0$. The recursion relation becomes

$$a_{k+1} = -\frac{1}{(2k+2)(2m+2k+2)}a_k = -\frac{1}{(k+1)(m+k+1)}a_k\frac{1}{2^2} \qquad (3.27)$$

These coefficients are to be inserted into (3.24) which we now write as

$$J_m(x) = \sum_{k=0}^{\infty} a_k x^{m+2k}$$

By convention, we write $a_0 = 1/m!2^m$ and therefore

$$\begin{aligned}
a_1 = \frac{(-1)}{(1)(m+1)}a_0\frac{1}{2^2} &= \frac{(-1)}{(1)(m+1)!}\frac{1}{2^{m+2}} \\
a_2 = \frac{(-1)}{(2)(m+2)}a_1\frac{1}{2^2} &= \frac{(-1)^2}{(2\cdot 1)(m+2)!}\frac{1}{2^{m+4}} \\
\text{that is} \quad a_k &= \frac{(-1)^k}{(k!)(m+k)!}\frac{1}{2^{m+2k}}
\end{aligned}$$

Finally, then, we arrive at the series expansion for the Bessel Function of integer order as

$$J_m(x) = \sum_{k=0}^{\infty} \frac{(-1)^k}{k!(m+k)!}\left(\frac{x}{2}\right)^{m+2k} \qquad m = 0, 1, 2, 3, \ldots \qquad (3.28)$$

Figure 3.9 shows on the top $J_0(x)$ using (3.28) summing over $k = 0, 1, \ldots, 10$. The function oscillates with decreasing amplitude, but diverges near $x = 9$. This is not surprising, given that the polynomial grows rapidly when the last term is proportional to x^{10}.

Of course, nobody ever uses (3.28) explicitly to calculate Bessel Functions. The theory of Bessel Functions provides a number of useful algorithms for calculating $J_m(x)$ to a high precision for any x, and any number of applications or software libraries give you access to these algorithms. The top plot of Figure 3.9 also shows $J_0(x)$ as calculated by some high precision technique. The bottom plots the first three integer order Bessel Functions $J_0(x)$, $J_1(x)$, and $J_2(x)$. Note how the behavior near $x = 0$ closely tracks x^m for $J_m(x)$.

3.6.3 SPHERICAL BESSEL FUNCTIONS

A very common (partial) differential equation in physics is the Helmholtz Equation, namely

$$\vec{\nabla}^2 u(\mathbf{r}) + k^2 u(\mathbf{r}) = 0 \qquad (3.29)$$

where $u(\mathbf{r})$ is a function in three-dimensional space of the coordinate $\mathbf{r}$. We will about the Laplacian operator $\vec{\nabla}^2$, and then the differential equations that include it,

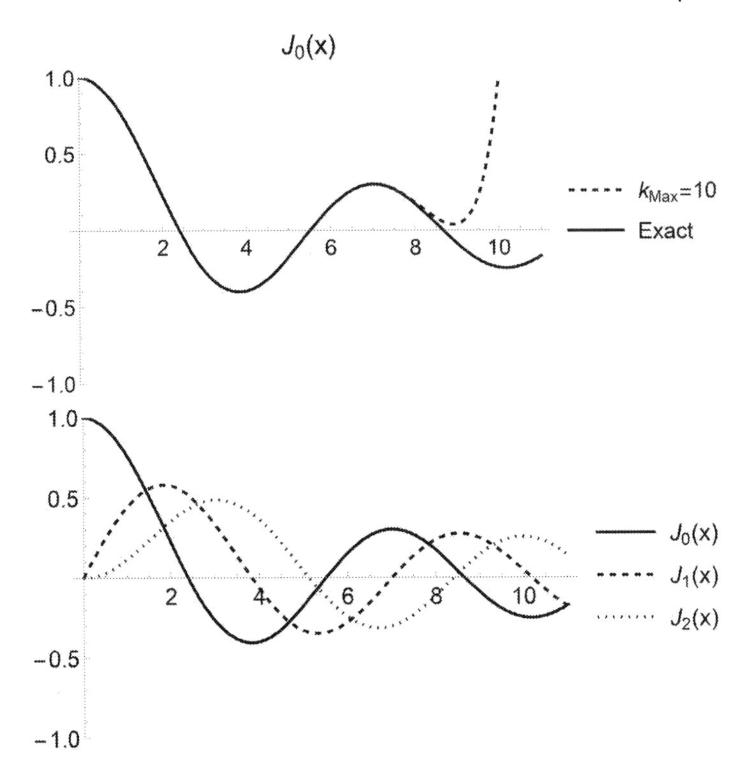

Figure 3.9 Plots of Bessel functions $J_m(x)$ for non-negative integer orders m. The top plot is for $m = 0$, where the dashed curve uses (3.28) summing up to $k = 10$, and the solid curve is the "exact" $J_0(x)$. The bottom plot shows the (exact) functions $J_0(x)$, $J_1(x)$, and $J_2(x)$.

in Section 4.2.4. For now, however, what is important to realize is the when we solve this equation in spherical coordinates, the ordinary differential equation

$$r^2R''(r) + 2rR'(r) + \left[k^2r^2 - \ell(\ell+1)\right]R(r) = 0 \tag{3.30}$$

where ℓ is a non-negative integer. This equation needs to be solved for $R(r)$, where $r = |\mathbf{r}|$. If we switch to a (dimensionless) variable $x = kr$, and write $R(r) = x^{-1/2}y(x)$, then

$$\begin{aligned}
R'(r) &= \frac{dR}{dr} = k\frac{d}{dx}\left[x^{-1/2}y(x)\right] = k\left[\frac{y'(x)}{x^{1/2}} - \frac{1}{2}\frac{y(x)}{x^{3/2}}\right] \\
2rR'(r) &= 2x^{1/2}y'(x) - \frac{y(x)}{x^{1/2}} \\
R''(r) &= \frac{d}{dr}R'(r) = k\frac{d}{dx}\left\{k\left[\frac{y'(x)}{x^{1/2}} - \frac{1}{2}\frac{y(x)}{x^{3/2}}\right]\right\} = k^2\left[\frac{y''(x)}{x^{1/2}} - \frac{y'(x)}{x^{3/2}} + \frac{3}{4}\frac{y(x)}{x^{5/2}}\right] \\
r^2R''(r) &= x^{3/2}y''(x) - x^{1/2}y'(x) + \frac{3}{4}\frac{y(x)}{x^{1/2}}
\end{aligned}$$

Inserting this into (3.30) and multiplying through by $x^{1/2}$ gives

$$x^2y''(x) + xy'(x) - \frac{1}{4}y(x) + \left[x^2 - \ell(\ell+1)\right]y(x) = 0$$

Now $\ell(\ell+1) + 1/4 = \ell^2 + \ell + 1/4 = (\ell+1/2)^2$ so we finally have

$$x^2y''(x) + xy'(x) + \left[x^2 - \left(\ell + \frac{1}{2}\right)^2\right]y(x) = 0 \tag{3.31}$$

which is a Bessel's Equation for $\nu = \ell + 1/2$. Therefore, solutions to (3.31) are of the form

$$y(x) = c_1 J_{\ell+1/2}(x) + c_2 Y_{\ell+1/2}(x)$$

Of course, physically, we are interested in $R(r) = (kr)^{-1/2}y(kr)$, so it is customary to define

$$j_\ell(x) \equiv \sqrt{\frac{\pi}{2x}}J_{\ell+1/2}(x) \tag{3.32a}$$

$$\text{and} \qquad n_\ell(x) \equiv \sqrt{\frac{\pi}{2x}}Y_{\ell+1/2}(x) \tag{3.32b}$$

These are known as the *Spherical Bessel Functions*. It is not very difficult to show that

$$j_0(x) = \frac{\sin x}{x} \qquad \text{and} \qquad n_0(x) = -\frac{\cos x}{x}$$

and that there is a recurrence relation for the higher orders, namely

$$f_\ell(x) = (-1)^\ell x^\ell \left(\frac{1}{x}\frac{d}{dx}\right)^\ell f_0(x)$$

where $f_\ell(x)$ can be either $j_\ell(x)$ or $n_\ell(x)$. It is interesting note that $j_\ell(x)$ and $n_\ell(x)$, unlike the $J_\nu(x)$, can be written in terms of sine and cosine functions.

Figure 3.10 plots $j_\ell(x)$and $n_\ell(x)$ for the three lowest values of ℓ. The $n_\ell(x)$ are singular for $x \to 0$, but you are much more likely to encounter the $j_\ell(x)$ in future physics course. The most important physics application I'm aware of, for the $n_\ell(x)$, have to do with scattering problems in Quantum Mechanics.

3.6.4 LEGENDRE POLYNOMIALS

In physical problems involving the Helmholtz Equation (3.29) where the system has spherical symmetry, that is $u(\mathbf{r}) = u(r)$, you encounter the Legendre Equation

$$(1-x^2)y''(x) - 2xy'(x) + \ell(\ell+1)y(x) = 0 \tag{3.33}$$

where $x = \cos\theta$ is defined in terms of the polar angle $0 \leq \theta \leq \pi$. For most physical problems of interest, ℓ is a non-negative integer, that is, $\ell = 0, 1, 2, \ldots$.

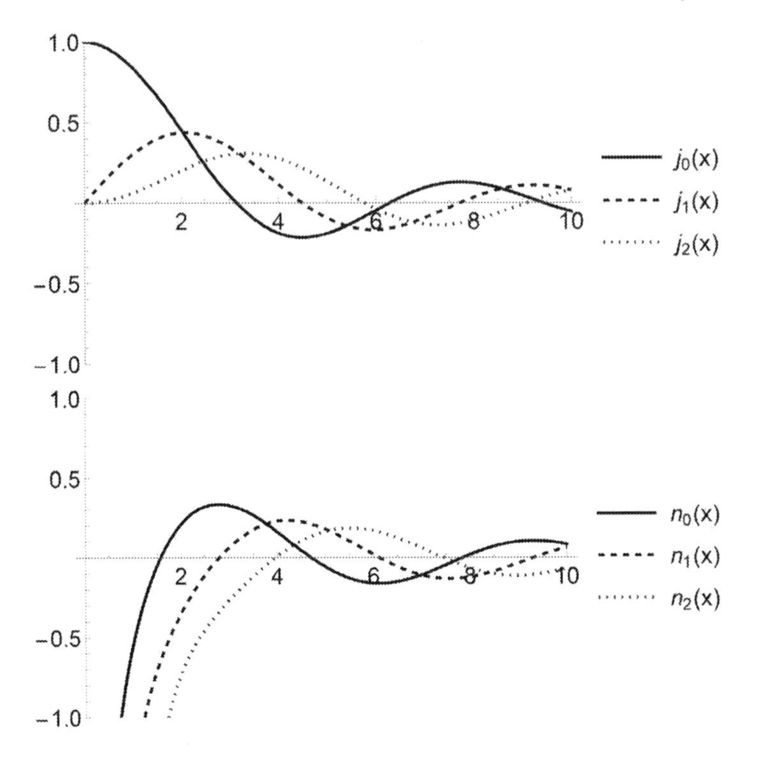

Figure 3.10 The Spherical Bessel Functions $j_\ell(x)$ (top) and $n_\ell(x)$ (bottom) for $\ell = 0, 1, 2$.

Legendre's Equation has regular singular points at $x = \pm 1$, but not at $x = 0$, so the simple series approach, setting

$$y = \sum_{n=0}^{\infty} a_n x^n$$

should work fine. Substituting this into (3.33) gives

$$\sum_{n=0}^{\infty} \left\{ n(n-1)a_n x^{n-2} + \left[-n(n-1) - 2n + \ell(\ell+1)\right] a_n x^n \right\} = 0$$

which results in the recursion relation

$$a_{n+2} = \frac{n(n-1) + 2n - \ell(\ell+1)}{(n+2)(n+1)} a_n = \frac{n(n+1) - \ell(\ell+1)}{(n+2)(n+1)} a_n$$

The free parameters are therefore a_0 and a_1, and these determine the two series, one with only even powers of x, and one with only odd powers.

Clearly, $a_{\ell+2} = 0$ for any given ℓ, so the solutions to (3.33) are polynomials of degree ℓ, called Legendre Polynomials $P_\ell(x)$. By convention, we set $P_\ell(1) = 1$, and

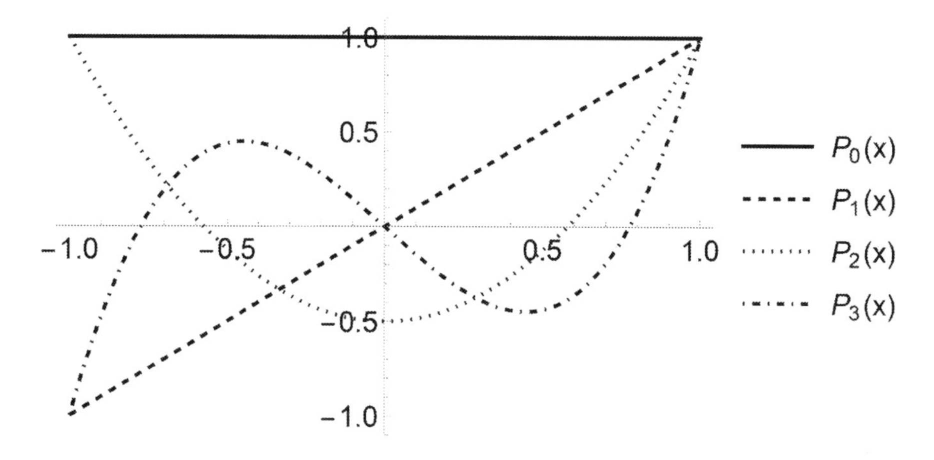

Figure 3.11 The first four Legendre Polynomials $P_0(x)$, $P_1(x)$, $P_2(x)$, and $P_3(x)$,

this fixes a_0 and a_1. The first several Legendre Polynomials are

$$\begin{aligned} P_0(x) &= 1 \\ P_1(x) &= x \\ P_2(x) &= \frac{1}{2}(3x^2-1) \\ P_3(x) &= \frac{1}{2}(5x^3-3x) \\ P_4(x) &= \frac{1}{8}(35x^4-30x^2+3) \\ P_5(x) &= \frac{1}{8}(63x^5-70x^3+15x) \end{aligned}$$

Figure 3.11 plots the Legendre Polynomials for $\ell=0$, $\ell=1$, $\ell=2$, and $\ell=3$.

The Legendre polynomials can also be written as

$$P_\ell(x) = \frac{1}{2^\ell \ell!}\frac{d^\ell}{dx^\ell}(x^2-1)^\ell \tag{3.34}$$

known as Rodrigues' Formula. This can be shown by direct substitution into (3.33). Using (3.34) it is possible to prove that

$$\int_{-1}^{1} P_\ell(x)P_\ell(x)\,dx = \frac{2}{2\ell+1} \tag{3.35}$$

by repeated integration by parts. In fact, it is not hard to show that

$$\int_{-1}^{1} P_\ell(x)P_\ell(x)\,dx = 0 \qquad \text{if} \qquad \ell \neq m \tag{3.36}$$

It is in this sense that we way that the Legendre polynomials are "orthogonal." (See Section 6.2.5.)

3.6.5 OTHER SPECIAL FUNCTIONS

We have only scratched the surface of "special" functions, defined in terms of the differential equations that they satisfy. Not only have we only barely talked about Bessel functions, spherical Bessel functions, and Legendre polynomials, and also the Hermite polynomials (in Section 3.5.2), there are very many other functions of this sort that are important in physics problems. I will try to give you here a brief glimpse of some of these.

A natural extension for problems with spherical symmetry gives you a differential equation for $x = \cos\theta$ with a new integer parameter m, namely the Associated Legendre Equation

$$(1-x^2)y''(x) - 2xy'(x) + \left[\ell(\ell+1) - \frac{m^2}{1-x^2}\right]y(x) = 0 \qquad (3.37)$$

The solutions to this equation are the Associated Legendre Functions $P_\ell^m(x)$, which clearly reduce to the $P_\ell(x)$ for $m = 0$. Note that the $P_\ell^m(x)$ are *not* polynomials.

When solving the hydrogen atom problem in Quantum Mechanics, you will need to solve for the radial dependence of the solution of the Schroödinger Equation. For this, you will encounter the Laguerre Equation, namely

$$xy''(x) + (1-x)y'(x) + ny(x) = 0 \qquad (3.38)$$

where n is a non-negative integer. The solutions to this equation are the Laguerre Polynomials. There are a lot of similarities with the Legendre Polynomials.

Kummer's Equation is a general second-order differential equation with "free" parameters a and b, where setting the parameters to different values gives you other well-known differential equations. Kummer's Equation is written as

$$xy''(x) + (b-x)y'(x) - ay(x) = 0 \qquad (3.39)$$

which obviously reduced to (3.38) for $b = 1$ and $a = -n$. One solution to (3.39) is

$$F(a;b;x) = 1 + \frac{a}{b}\frac{x}{1!} + \frac{a(a+1)}{b(b+1)}\frac{x^2}{2!} + \cdots$$

which is called the *confluent hypergeometric function*. You can begin to see how the Laguerre polynomials could be derived from this formula, but this is only the beginning of a long story.

3.7 COUPLED DIFFERENTIAL EQUATIONS

It is possible, and physically likely, that a system will be governed by more than one differential equation, for more than one dependent variable, and that these dependent variables appear in more than one of the equations. In this case, we say that the differential equations are *coupled*. In general, for these cases, we will have to be clever in order to find a solution, even if the equations are linear and homogeneous.

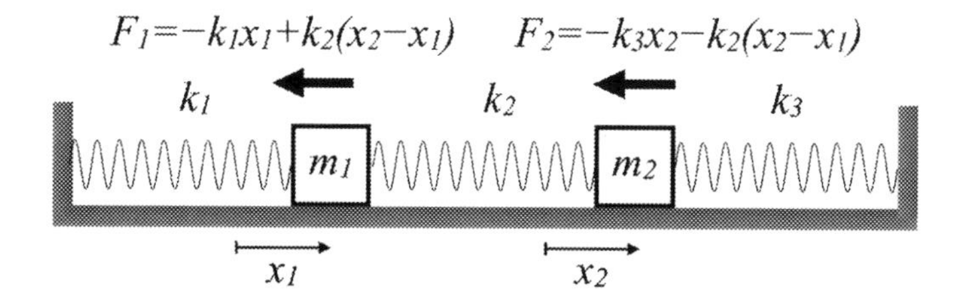

Figure 3.12 A coupled harmonic oscillator with two masses and three springs.

Remember, for the purposes of this course, at least, we are guided by existence and uniqueness.

Rather than try to treat coupled differential equations in general, we will take the opportunity to solve a specific problem, namely the motion of coupled simple harmonic oscillators. We will see that generalizing the ansatz we used for the oscillator will set us in the right direction. It will also provide hints to the concept of *eigenvalues*, which we will cover more thoroughly in Section 6.4.

3.7.1 COUPLED SIMPLE HARMONIC OSCILLATORS

Figure 3.12 shows a prototype of coupled linear second-order differential equations. Two masses m_1 and m_2 are each attached by springs to a fixed wall. They are also connected by a "coupling" spring, and the force on each mass due to the coupling spring depends on the positions of each of the masses. Note that the force the coupling spring exerts on m_1 is equal and opposite to the force it exerts on m_2. Newton's Second Law, applied separately to the two masses, becomes

$$\begin{aligned} m_1\ddot{x}_1 = F_1 &= -k_1x_1 + k_2(x_2 - x_1) \\ m_2\ddot{x}_2 = F_2 &= -k_3x_2 - k_2(x_2 - x_1) \end{aligned}$$

If you have trouble seeing the sign on the coupling force, just think about what happens if m_2 is extended more than m_1 in which case the spring wants to compress, moving m_1 to the right and m_2 to the left.

At this point we will make the problem even more specific, and set $m_1 = m_2 = m$ and $k_1 = k_2 = k_3$. Defining $\omega_0^2 = k/m$, we get the differential equations

$$x_1''(t) + 2\omega_0^2 x_1(t) - \omega_0^2 x_2(t) = 0 \tag{3.40a}$$

$$x_2''(t) + 2\omega_0^2 x_2(t) - \omega_0^2 x_1(t) = 0 \tag{3.40b}$$

These differential equations are clearly "coupled." The equation for $x_1(t)$ depends on the function $x_2(t)$, and the equation for $x_2(t)$ depends on the function $x_1(t)$.

We will approach the solution using the ansatz

$$x_1(t) = a_1 e^{i\omega t} \qquad \text{and} \qquad x_2(t) = a_2 e^{i\omega t}$$

and look to see what we can learn about ω, a_1, and a_2. This ansatz would seem to imply that both masses oscillate at the same frequency ω. However, let's plow forward and see what happens. Inserting this ansatz into (3.40) we have

$$\begin{aligned} -\omega^2 a_1 + 2\omega_0^2 a_1 - \omega_0^2 a_2 &= 0 \\ -\omega^2 a_2 + 2\omega_0^2 a_2 - \omega_0^2 a_1 &= 0 \end{aligned}$$

Let me rewrite these equations in a suggestive form, and that will become more clear to you when we cover systems of linear equations in Section 6.3.9. We have

$$(2\omega_0^2 - \omega^2)a_1 - \omega_0^2 a_2 = 0 \tag{3.41a}$$

$$-\omega_0^2 a_1 + (2\omega_0^2 - \omega^2)a_2 = 0 \tag{3.41b}$$

One obvious solution to these equations is $a_1 = a_2 = 0$, but that just means that neither mass ever moves. There would also be no way to accommodate arbitrary initial conditions for position and velocity for each of the masses. Our notions of existence and uniqueness tell us that there has to be another way to solve these equations.

Remember that we have ω^2 to play with. If ω^2 was set to some value that made both equations the *same* equation, then all we could get out of this would be constraints on the ratio of a_1/a_2, in which case we can use initial conditions to solve the rest of the problem. The two equations (3.41) become one equation if the coefficients of a_1 and a_2 are in the same ratio, that is

$$\frac{2\omega_0^2 - \omega^2}{-\omega_0^2} = \frac{-\omega_0^2}{2\omega_0^2 - \omega^2}$$

This equation is simple to solve for ω^2. We get

$$(2\omega_0^2 - \omega^2)^2 = \omega_0^4 \qquad \text{so} \qquad \omega^2 = 2\omega_0^2 \pm \omega_0^2 = \omega_0^2,\, 3\omega_0^2$$

Indeed, there are two frequencies ω for which the two masses can oscillate together. Of course, the real solution can be of any linear combination of these solutions, each of which is proportional to $e^{\pm i\omega t}$. There is an important constraint on these solutions, though, namely the ratio of a_1 and a_2 that comes from (3.41). In the case where $\omega^2 = \omega_0^2$, we find

$$\omega_0^2 a_1 - \omega_0^2 a_2 = 0 \qquad \text{so} \qquad a_2 = a_1 \qquad \text{where} \qquad \omega^2 = \omega_0^2$$

while for $\omega^2 = 3\omega_0^2$,

$$-\omega_0^2 a_1 - \omega_0^2 a_2 = 0 \qquad \text{so} \qquad a_2 = -a_1 \qquad \text{where} \qquad \omega^2 = 3\omega_0^2$$

In other words, if the two masses are going to oscillate at the same frequency, then they either follow each other with the same amplitude and phase (at frequency $\omega = \omega_0$) or they follow each other with the same amplitude but 180° out of phase (at frequency $\omega = \sqrt{3}\omega_0$).

The general solution can now be written in terms of the four constants a, b, c, and d as

$$x_1(t) = ae^{i\omega_0 t} + be^{-i\omega_0 t} + ce^{i\sqrt{3}\omega_0 t} + de^{-i\sqrt{3}\omega_0 t} \quad (3.42a)$$

$$x_2(t) = ae^{i\omega_0 t} + be^{-i\omega_0 t} - ce^{i\sqrt{3}\omega_0 t} - de^{-i\sqrt{3}\omega_0 t} \quad (3.42b)$$

The constants are determined from the initial conditions on $x_1(0)$, $\dot{x}_1(0)$, $x_2(0)$, and $\dot{x}_2(0)$. I urge you to take the time to insert (3.42) into (3.40) and confirm that these are in fact solutions for any a, b, c, and d.

Notice that, regardless of the initial conditions, the linear combination $x_+(t) \equiv x_1(t) + x_2(t)$ oscillates at the frequency ω_0, while the linear combination $x_-(t) \equiv x_1(t) - x_2(t)$ oscillates at the frequency $\sqrt{3}\omega_0$. In Chapter 6, we will learn to call ω_0^2 and $3\omega_0^2$ *eigenvalues*, and the solutions $x_\pm(t)$ will be components of *eigenvectors*. Somewhat more colloquially, we'll call the two kinds of motion associated with $x_\pm(t)$ "eigenmodes."

It is straightforward, but a little tedious, to take the four equations

$$x_1(0) = x_{1_0} \qquad \dot{x}_1(0) = v_{1_0} \qquad x_2(0) = x_{2_0} \qquad \dot{x}_2(0) = v_{2_0}$$

and solve for a, b, c, and d, so I won't bother to write out the solution. (Of course, you can just feed this into some mathematical symbolic manipulations application and let it solve the equations for you.) In the end, you find

$$a = \frac{1}{4}x_{1_0} + \frac{1}{4}x_{2_0} - \frac{i}{4\omega_0}v_{1_0} - \frac{i}{4\omega_0}v_{2_0} \quad (3.43a)$$

$$b = \frac{1}{4}x_{1_0} + \frac{1}{4}x_{2_0} + \frac{i}{4\omega_0}v_{1_0} + \frac{i}{4\omega_0}v_{2_0} \quad (3.43b)$$

$$c = \frac{1}{4}x_{1_0} - \frac{1}{4}x_{2_0} - \frac{i}{4\sqrt{3}\omega_0}v_{1_0} + \frac{i}{4\sqrt{3}\omega_0}v_{2_0} \quad (3.43c)$$

$$d = \frac{1}{4}x_{1_0} - \frac{1}{4}x_{2_0} + \frac{i}{4\sqrt{3}\omega_0}v_{1_0} - \frac{i}{4\sqrt{3}\omega_0}v_{2_0} \quad (3.43d)$$

Figure 3.13 shows the motions $x_1(t)$ and $x_2(t)$ for initial conditions $x_{1_0} = 1$ and $x_{2_0} = v_{1_0} = v_{2_0} = 0$. Time is measured here in units of $2\pi/\omega_0$. The motion is periodic, although not purely harmonic. Indeed, these motions are the mixture of two frequencies, namely ω_0 and $\sqrt{3}\omega_0$.

Figure 3.13 also plots $x_+(t)$ and $x_-(t)$, which now show clearly the two fundamental frequencies of this two-body coupled system.

It is not hard to set the two masses in motion so that they oscillate each at the same frequency. All we need to do is set the initial conditions to respect the relationships between the two amplitudes for the frequency eigenvalue in question. This is shown in Figure 3.14 where we set $x_2(0) = x_1(0)$ on the left, resulting in both oscillating together at frequency ω_0 – note that the period of oscillation is clearly equal to unity (in units of $2\pi/\omega_0$). The figure also shows that when $x_2(0) = -x_1(0)$, the two masses again oscillate together, but now at frequency $\sqrt{3}\omega_0$.

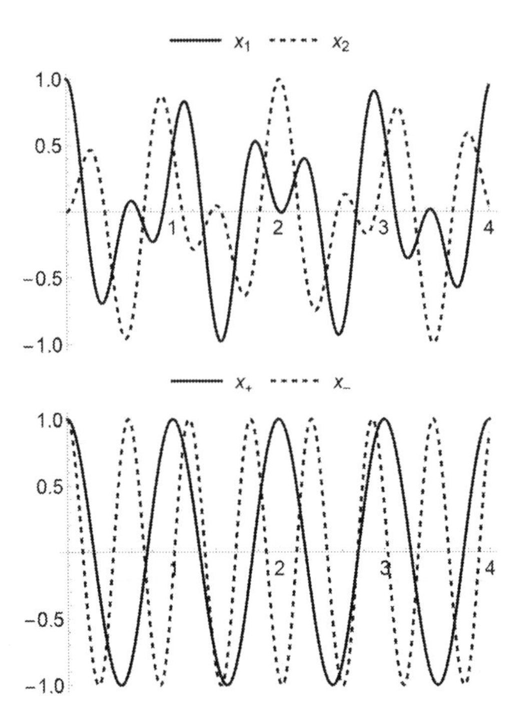

Figure 3.13 Plots of the motions of two coupled masses as shown in Figure 3.12 and described by (3.42) and (3.43) for initial conditions $x_1(0) = 1$ and $x_2(0) = \dot{x}_1(0) = \dot{x}_2(0) = 0$. The top plots the positions of the individual masses $x_1(t)$ and $x_2(t)$ as a function of time in units of $2\pi/\omega_0$. The bottom plots the linear combinations $x_+(t) \equiv x_1(t) + x_2(t)$ and $x_-(t) \equiv x_1(t) - x_2(t)$. The lesson here is that, for these initial conditions, the individual masses do not oscillate with a single frequency, but their center of mass and relative positions do.

EXERCISES

3.1 *Find the solution $y(x)$ to the first-order linear differential equation*

$$y'(x) = x + y(x) \qquad \text{with} \qquad y(0) = 0$$

3.2 *Find the solution of the differential equation $dx/dt = x^2$ where x is position and t is time, and where $x(0) = a$ with $a > 0$. For what range of times is your solution valid? (Be careful! This is a trick question.)*

3.3 *The pressure $P(T)$ along a liquid-gas phase boundary on a pressure vs temperature (T) diagram is the solution to the differential equation*

$$\frac{dP}{P} = k\frac{dT}{T^2}$$

where k is a a constant. If the pressure is P_0 when the temperature is T_0, find the function $P(T)$.

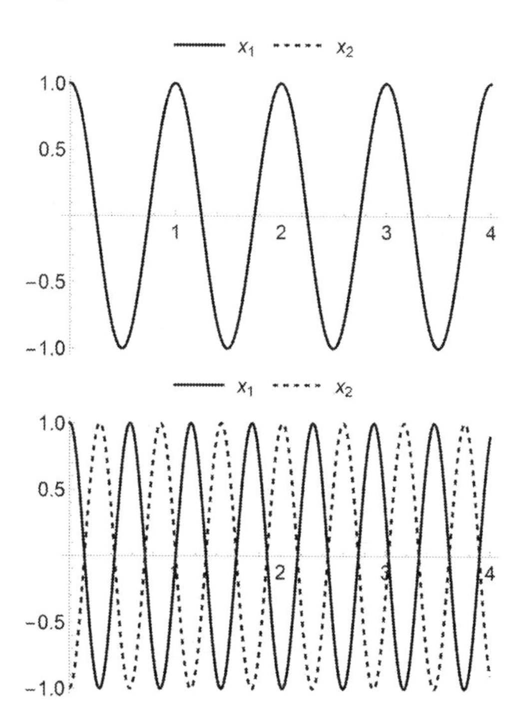

Figure 3.14 The motions $x_1(t)$ and $x_2(t)$ for initial condition variables $x_{1_0} = 1$ and $x_{2_0} = 1$ (top) and $x_{1_0} = 1$ and $x_{2_0} = -1$ (bottom). (The initial velocities are still $v_{1_0} = v_{2_0} = 0$.) In this case, the initial conditions correspond to non-oscillating relative positions (top) or center of mass (bottom) coordinates. Note that $x_1(t)$ and $x_2(t)$ are identical when the initial conditions have equal displacements for the two masses.

3.4 *Find the solution $y(x)$ to the second-order linear differential equation*

$$y''(x) + y(x) = \sin(2x) \qquad \text{with} \qquad y(0) = 0, y'(0) = 0$$

3.5 *A function $y = y(x)$ obeys the second-order linear differential equation*

$$y'' + f(x)y' + g(x)y = 0$$

where $f(x)$ and $g(x)$ are arbitrary functions. Show that if $y_1(x)$ and $y_2(x)$ both solve the differential equation, then the linear combination $y_3 = ay_1(x) + by_2(x)$ also solves the equation, where a and b are arbitrary constants. This is called the Principle of Superposition.

3.6 *Use the "integrating factor" approach to find the solution $y(x)$ to the differential equation*

$$\frac{dy}{dx} - y = 2xe^{2x}$$

subject to the boundary condition $y(0) = 1$.

3.7 *Assume that a spherical raindrop evaporates at a rate proportional to its surface area. If its radius is initially 3 mm, and one hour later its radius is 2 mm, find the radius of the raindrop at any time t.*

3.8 *Given the second-order linear differential equation*

$$x^2 \frac{d^2y}{dx^2} + 2x\frac{dy}{dx} - 1 = 0 \qquad \text{where} \qquad x > 0$$

Find the general solution for the function $y(x)$. (The general solution is a solution with two arbitrary constants.) You can do this by first converting the equation to first order, and then integrating the result.

3.9 *Find the unique solution $y(x)$ to the differential equation and boundary conditions*

$$\frac{d^2y}{dx^2} + \frac{dy}{dx} - 2y = 2x \qquad \text{with} \qquad y(0) = 0 \qquad \text{and} \qquad y'(0) = 1$$

You can do this by using the ansatz $y(x) = e^{\alpha x}$ to find the general solution to the homogeneous equation, and then make a good guess to find a particular solution to the complete equation. Given this, use the boundary conditions to find the two otherwise arbitrary constants from the general solution.

3.10 *An object of mass m falls from rest some distance above the Earth's surface. It is subject to a drag force av^2 proportional to its velocity. Find its velocity $v(t)$, and check that your answer is dimensionally correct. Then check that you get the correct behavior for both short and long times. I suggest that that you choose a coordinate system where "up" is positive.*

3.11 *A capacitor C is connected in series with a resistor R as shown below:*

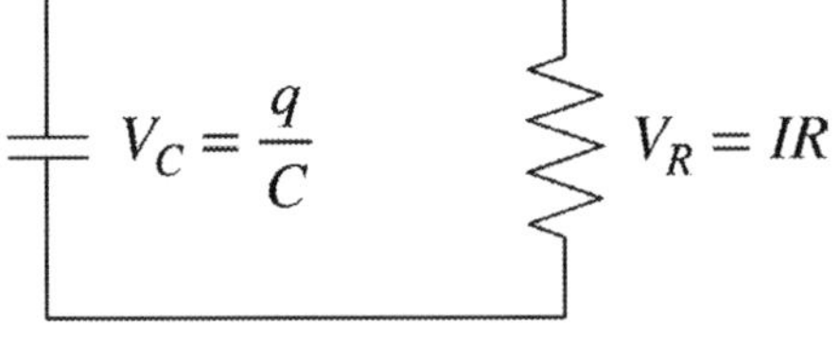

$$V_C = \frac{q}{C} \qquad V_R = IR$$

The potential difference across the capacitor is $V_C = q/C$ where q is the charge stored on the capacitor. The potential difference across the resistor is $V_R = IR$ where $I = dq/dt$ is the current through the resistor. If the initial charge on the capacitor is q_0, find $q(t)$ as a function of time.

3.12 *A capacitor C is connected in series with an inductor L as shown below:*

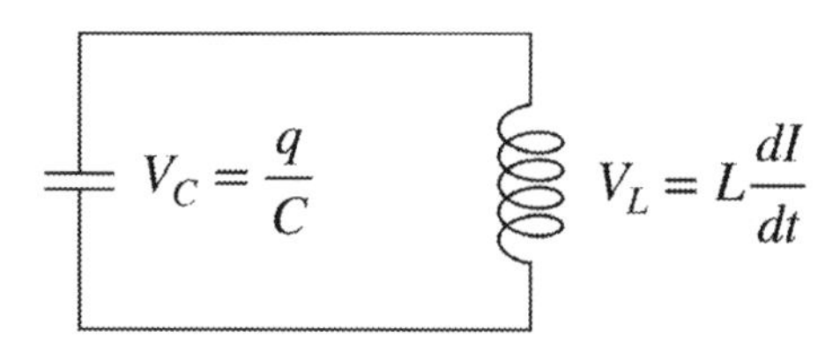

The potential difference across the capacitor is $V_C = q/C$ where q is the charge stored on the capacitor. The potential difference across the inductor is $V_L = LdI/dt$ where $I = dq/dt$ is the current through the inductor. If the initial charge on the capacitor is q_0, and the initial current is zero, find $q(t)$ as a function of time.

3.13 *An object of mass m moves in one dimension x and is subjected to a force $F(t) = F_0 e^{-\alpha t}$ where F_0 and α are positive constants. Write the second-order differential equation that governs the motion, and solve it to find the position $x(t)$ as a function of time if the particle starts from rest at $x = 0$.*

3.14 *Make some sketches of the motion of $x(t)$ for the damped oscillator, similar to Figure 3.3 and 3.4. Plot against time in units of the fundamental period of the undamped oscillator. Plot the following cases:*

(a) $\beta = 0.05\omega_0$, $x_0 = 5$, $v_0 = 2\omega_0$
(b) $\beta = 1.5\omega_0$, $x_0 = 1$, $v_0 = -2\omega_0$
(c) $\beta = \omega_0$, $x_0 = 1$, $v_0 = -2\omega_0$

3.15 *Reproduce the plots in Figure 3.6, that is $x(t)$ for a driven damped oscillator with $\beta = 0.05\omega_0$, $\gamma = 10$, and with initial conditions $x(0) = \dot{x}(0) = 0$. The three plots are for $\omega = 0.5\omega_0$, $\omega = \omega_0$, and $\omega = 1.5\omega_0$. Time is plotted in units of the fundamental period of the undamped oscillator. (This is a nice demonstration of resonance.)*

3.16 *Referring to the terminology used in the text, we saw that if $\beta > \omega_0$ ("over damping"), then the solution to the damped oscillator is the sum of two exponential functions, no matter how close β is to ω_0. However, if $\beta = \omega_0$ ("critical damping"), the solution magically turns into a single exponential dependence. Write $\omega_0^2 = \beta^2(1 - \varepsilon^2)$ and show that for $\varepsilon \ll 1$ the over damped solution turns into the critically damped solution.*

3.17 *Use the series approach to find the solutions for $y'' = y(x)$ and show that the result is the same as the series expansion for $y(x) = c_1 \cosh(kx) + c_2 \sinh(kx)$. How would you define constants a_1 and a_2 in terms of c_1 and c_2 so that the solution is $y(x) = a_1 e^x + a_2 e^{-x}$?*

3.18 *Find general solution to $y''(x) = xy(x)$ using series approach. Show that there can be no term in the series proportional to x^2, and that the recursion relations relate every third term of the expansion. Separate the two solutions you and explicitly indicate the constants of integration. Write out the first ten or so nonzero terms of each of the two solutions, and plot them. (Don't go too far in $\pm x$ so that you over run the range afforded by the number of terms you calculated in the expansion!) Note the difference in behavior for $x < 0$ and $x > 0$.*

3.19 *Consider the Euler Equations (3.22).*

(a) Use the Wronskian to show that the two solutions for $(\alpha - 1)^2 > 4\beta^2$ are linearly independent for all $x > 0$. Recall that the Wronskian for two solutions $y_1(x)$ and $y_2(x)$ is $W[y_1(x), y_2(x)] = y_1(x)y_2'(x) - y_2(x)y_1'(x)$.

(b) For the case $(\alpha - 1)^2 = 4\beta^2$, find a second solution and again show that the two solutions are linearly independent for all $x > 0$. Hint: Try the approach that worked for a linear second-order equation with constant coefficients.

3.20 *Show that, for a Bessel Function $J_m(x)$ for integer order m, $J_{-m}(x) = (-1)^m J_m(x)$. You can use the Γ–Function to interpret $n!$ for $n < 0$. Explain why this means that $y(x) = c_1 J_m(x) + c_2 J_{-m}(x)$ cannot be the general solution to Bessel's Equation for $m \in \mathbb{Z}$.*

3.21 *Show by explicit substitution that the Spherical Bessel Function $j_0(x) = \sin(x)/x$ of order zero, where $x = kr$, solves the $\ell = 0$ radial dependence of the Helmholtz Equation*

$$r^2 R''(r) + 2rR'(r) + k^2 r^2 R(r) = 0$$

3.22 *Use Rodrigues' Formula to derive the first three Legendre Polynomials $P_0(x)$, $P_1(x)$, and $P_2(x)$, and compare to the results given in the text.*

3.23 *Use the orthogonality of the Legendre Polynomials to derive an infinite series that gives $f(x) = e^x$ over the domain $-1 \leq x \leq 1$ in the form*

$$f(x) = \sum_{\ell=0}^{\infty} A_\ell P_\ell(x)$$

Plot your result for some number of terms of the series and compare to $f(x)$.

3.24 *A mass m moves in one dimension $x(t)$ connected to a spring with stiffness k, and is driven by a force term $F = ma\cos\omega t$ where a and ω are constants. Write down and solve the differential equation for $x(t)$ in terms of $\omega_0^2 \equiv k/m$ for the initial conditions $x(0) = \dot{x}(0) = 0$. Use a trigonometric identity to cast your solution in the form of a single product of sines. Write $\omega_0 - \omega = \varepsilon$ with $|\varepsilon| \ll \omega_0$, and describe the motion for short times $t \ll 1/|\varepsilon|$ (but $t \gg 1/\omega_0$) and long times $t \gg 1/|\varepsilon|$.*

3.25 *An mechanical oscillator has position $x(t)$ governed by the equations*

$$\ddot{x}(t) + 2\dot{x}(t) + 5x(t) = e^{-t}\cos(3t) \qquad x(0) = 0 \qquad \dot{x}(0) = 0$$

Find the motion $x(t)$ and plot it for $0 \le t \le 2\pi$.

3.26 *The diagram below is of an electrical circuit with a resistor R, capacitor C, inductor L, and an AC voltage source $V(t)$ connected in series:*

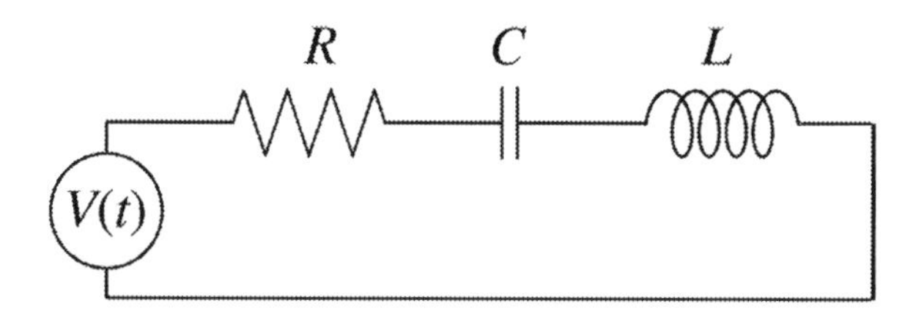

The voltage drop across the capacitor is q/C where $q(t)$ is the charge on the capacitor, the voltage drop across the resistor is iR where $i = dq/dt$ is the current in the circuit, and the voltage drop across the inductor is Ldi/dt. Kirchoff's Law says that the sum of all voltage drops around a closed path must be zero. If $V(t) = -V_0\cos(\omega t)$, then find $q(t)$ assuming that $q(0) = 0$ and $i(0) = 0$.

3.27 *Find the general solution $y(x)$ for the differential equation $y''(x) = y$ using the series solution approach, about $x = 0$, written as a linear combination of two separate infinite series. Show that that two series are in fact those for* $\cosh(x)$ *and* $\sinh(x)$.

3.28 *Find a series solution for $y(x)$ about $x = 0$ for the differential equation*

$$y'' - 2xy' + \lambda y = 0$$

in terms of two independent series solutions $y_0(x)$ and $y_1(x)$. For what values of λ is the solution a polynomial? Find the polynomial solution for $\lambda = 4$.

3.29 *Two masses and three springs are arranged as shown here:*

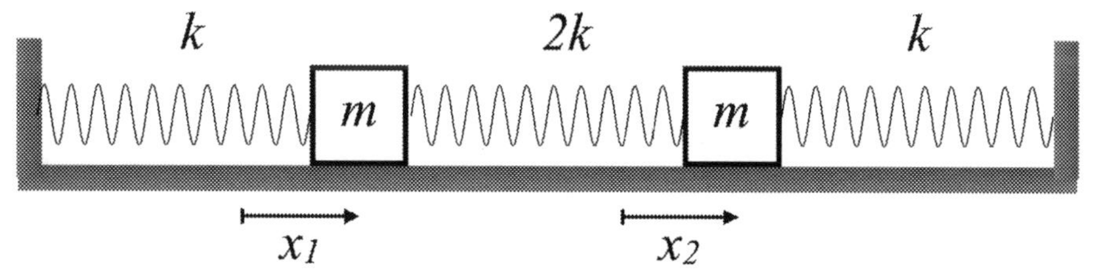

Find the eigenfrequencies for the two eigenmodes of oscillation, and determine the relative amplitudes of oscillation of the two masses for each mode.

3.30 *Two masses 3m and 2m are connected to two identical springs as shown:*

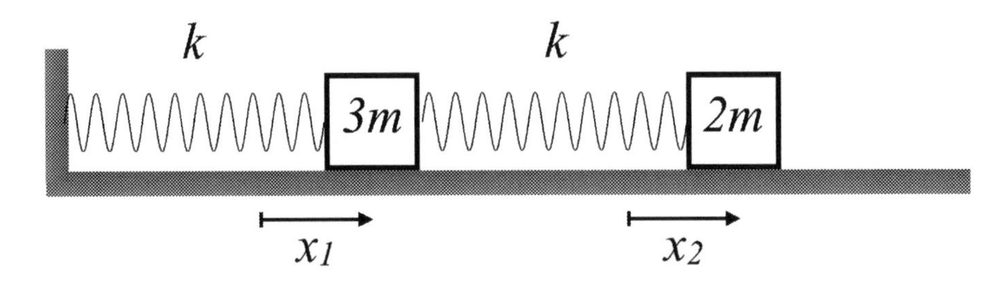

The masses are free to move horizontally and one spring is attached to a fixed wall.

(a) *Write down Newton's Second Law for each of the two masses.*
(b) *Find the eigenfrequencies and describe the motion of the two eigenmodes.*
(c) *Write $x_1(t)$ and $x_2(t)$ in terms of four arbitrary constants a, b, c, and d.*
(d) *Make a plot of $x_1(t)$ and $x_2(t)$ subject to the initial conditions $x_1(0) = 1$, and $x_2(0) = \dot{x}_1(0) = \dot{x}_2(0) = 0$.*
(e) *For the example in class, we found that the two combinations $x_\pm(t) = x_1(t) \pm x_2(t)$ oscillated with the two eigenfrequencies. What linear combinations $x_A(t)$ and $x_B(t)$ of $x_1(t)$ and $x_2(t)$ oscillate with the eigenfrequencies in this case? The answer should be clear from (c) above. Plot $x_A(t)$ and $x_B(t)$ and show they they oscillate with the correct frequencies.*

3.31 *Starting from the infinite power series expression for the Bessel function $J_m(x)$, where m is an integer, prove that*

$$\frac{d}{dx}\left[x^m J_m(x)\right] = x^m J_{m-1}(x)$$

3.32 *Prove that the Legendre polynomials are "orthogonal," that is $\int_{-1}^{1} P_\ell(x)P_m(x)\,dx = 0$ if $\ell \neq m$. You can do this by writing down the differential equation for $P_\ell(x)$ and multiplying through by $P_m(x)$. Then create a second equation by reversing the indices, subtract the two equations and then integrate.*

3.33 *It is possible to prove that $\int_{-1}^{1} P_\ell(x)P_\ell(x)\,dx = 2/(2\ell+1)$. (But we won't try to do that now.) Use this, along with the orthogonality of Legendre polynomials, to find an expression for the coefficients a_n in the expansion*

$$f(x) = \sum_{m=0}^{\infty} a_m P_m(x)$$

where $f(x)$ is defined for $-1 \leq x \leq 1$. You can do this by multiplying both sides of this expression by $P_\ell(x)$ and integrating. Now use this find the first few nonzero coefficients for $f(x) = \sin(\pi x)$ and make a plot of the expansion compared to $f(x)$.

3.34 *Two identical mass hang vertically under their own weight from two identical springs from a fixed point on the ceiling, as shown in the figure on the right. Find the two normal frequencies and describe the amplitudes of the two normal modes.*

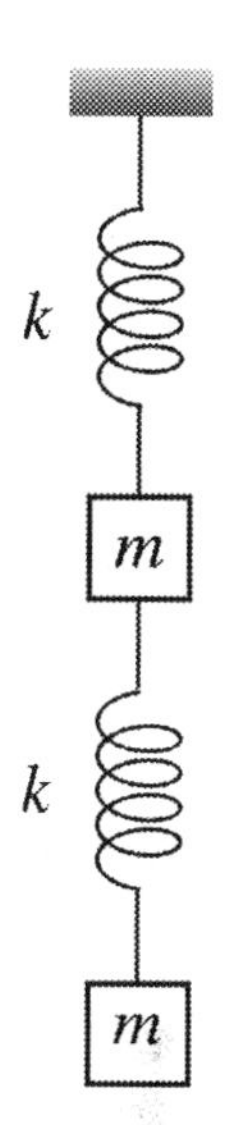

3.35 *Three identical capacitors C are connected to two identical inductors L as shown in the following figure. Find two coupled differential equations for $q_1(t)$ and $q_2(t)$ and find the normal mode frequencies. Analyze the problem by equating the potential differences for legs 1, 2, and 3 between nodes A and B. Use the sign convention shown for the currents in each of the three legs which implies that $i_1 + i_2 + i_3 = 0$. You can assume the charges are all zero when the currents are all zero.*

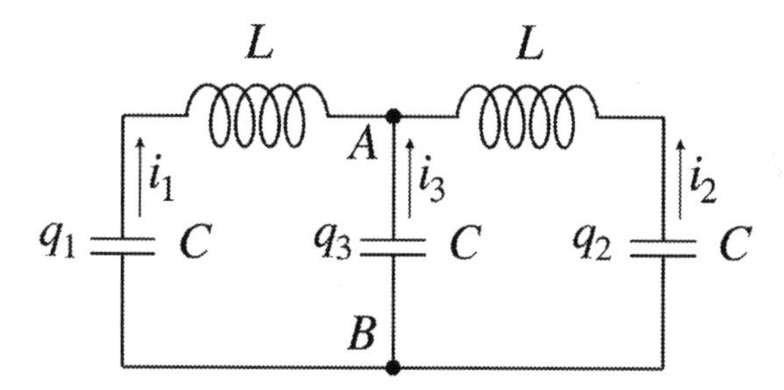

4 Vector Calculus and Partial Differential Equations

This chapter focuses on *vectors* as entities which describe quantities in ordinary three-dimensional space. We'll talk a little bit about generalizing what we mean by a "vector" in Section 4.1.5, but will take this generalization much further in Chapter 6.

4.1 VECTORS AS SPATIAL VARIABLES

Your first introduction to a "vector" was probably something written like $\vec{r}$ which stood for three (real number) values, likely called x, y, and z, and which located a point in three-dimensional space. A very simple form might have been $\vec{r} = (x, y, z)$, and this tells the position of the point by marking off the distance on the x-, y-, and z-axes. However, we will be more sophisticated than this, and write

$$\vec{r} = x\hat{\imath} + y\hat{\jmath} + z\hat{k} \tag{4.1}$$

where $\hat{\imath}$, $\hat{\jmath}$, and $\hat{k}$ are *unit vectors* in the x-, y-, and z-directions. I'll be more precise about what I mean by "unit vector" when we talk about the *inner product* in Section 4.1.2. For now, though, just think of $\hat{\imath}$, $\hat{\jmath}$, and $\hat{k}$ as dimensionless quantities of magnitude unity, but with a direction that is their respective axis.

Oftentimes, we are concerned only with vectors in a plane. In this case, the "z-component" of the vector is irrelevant for the problem at hand. Typically, we simply ignore it, although we may at some point need to refer to "the direction perpendicular to the plane" of whatever are the more relevant variables.

More generally, a vector $\vec{A}$ will be regarded as an element of $\mathbb{R}^2$ or $\mathbb{R}^3$ which represents some physical quantity in two-dimensional or three-dimensional space. Equation (4.1) is just one example. In fact, if it represents the physical location of some object that can move with time, then another obvious vector is the velocity, namely

$$\vec{v}(t) = \frac{d\vec{r}}{dt} = \frac{dx}{dt}\hat{\imath} + \frac{dy}{dt}\hat{\jmath} + \frac{dz}{dt}\hat{k}$$

Notice that I slipped it by you here that the unit vectors themselves do not change with time, which is true for this coordinate system, but not for some others. Of course, the next step would be to define the acceleration vector $\vec{a} = d\vec{v}/dt$. In general, we denote the components of a vector $\vec{A}$ as A_x, A_y, and A_z, and write

$$\vec{A} = A_x\hat{\imath} + A_y\hat{\jmath} + A_z\hat{k} \tag{4.2}$$

The *magnitude*, or "length" of a vector $\vec{A} = A_x\hat{\imath} + A_y\hat{\jmath} + A_z\hat{k}$ is

$$|\vec{A}| = \sqrt{A_x^2 + A_y^2 + A_z^2} \tag{4.3}$$

DOI: 10.1201/9781003355656-4

which of course is a positive real number. It makes sense, therefore, to borrow the notation $|\vec{A}|$ from the "magnitude" or a real or complex number. Very often, we will just write $A = |\vec{A}|$ if the context is clear. We will see a more formal way to define the magnitude of a vector when we study inner products in Section 4.1.2.

It happens often in physics that a vector $\vec{A}$ represents some physical quantity that could have different values at different points $\vec{r}$ in space, or even at different times. In other words, we have $\vec{A} = \vec{A}(\vec{r}) = \vec{A}(x,y,z)$ or $\vec{A} = \vec{A}(\vec{r},t) = \vec{A}(x,y,z,t)$. In this case, we refer to $\vec{A}$ as a *vector field*. Probably the first examples that come to mind are the electric field $\vec{E}$ and magnetic field $\vec{B}$, but there are many other examples, including many that do not come from electromagnetism. For example, a compressible fluid will have a "velocity field" $\vec{v} = \vec{v}(\vec{r},t)$ that would be governed by the theory of fluid mechanics.

We could also have a so-called "scalar field" which depends on position $\vec{r}$ (and possibly time t), although the use of the term "scalar" needs some consideration that we'll deal with later. We will soon be discussing "vector differential operators" which can turn a scalar field into a vector field, in exactly the same way that the static electric field $\vec{E}(\vec{r})$ can be derived from a static electric potential $\Phi(\vec{r})$.

We have been locating position in this section by representing $\vec{r}$ by the real numbers x, y, and z. These are referred to as "Cartesian coordinates." (I think this is because they were invented by the philosopher and mathematician Rene Descartes.) Likewise, $\hat{\imath}$, $\hat{\jmath}$, and $\hat{k}$ are "Cartesian unit vectors." However, there are other ways to locate a point in two- or three-dimensional space, namely be identifying the distance from the origin and then using one or two angles to tell the direction of the point with respect to the x-, y-, and z-axes. We will discuss these in Section 4.1.3.

First, however, we will say a few words about rotations, just to put the notion of vectors into the context that we'll eventually use to more precisely define them. Then we'll discuss two important geometrical concepts of vectors.

4.1.1 AXIS ROTATIONS

It doesn't matter, of course, what direction we pick when we define our Cartesian coordinate system. If the problem has some specific direction in it, for example the direction of an electric or magnetic field, then we often choose that direction to define one of the axes, typically the z-axis. In that case, the x-axis is chosen in some direction perpendicular to z, and this defines y in a way that makes the coordinate system "right handed." (This will be more precisely defined in Section 4.1.2.)

Regardless of our choice, however, we can always do physics in some set of axes (x',y',z') that are rotated with respect to (x,y,z). This rotation can be specified in any number of ways, but will typically involve three angles. (In rigid body classical mechanics, and in the theory of angular momentum in quantum mechanics, these three angles are called *Euler Angles*.) When we study matrix operations in Section 6.3 we will see that matrices with certain properties are a handy way to describe rotations, and these matrices form a *group*.

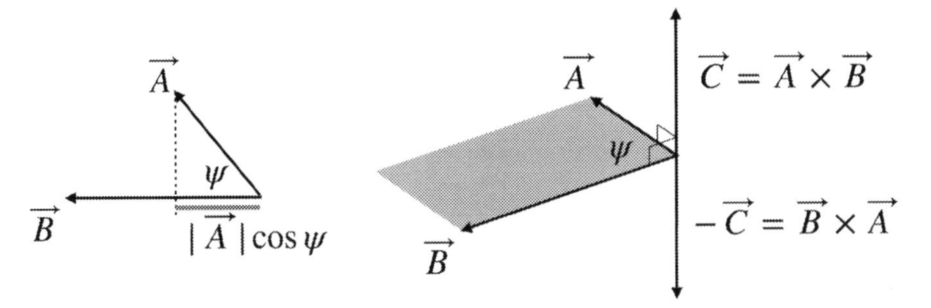

Figure 4.1 Geometric interpretations of the dot product $\vec{A} \cdot \vec{B}$ (left) and the cross product $\vec{A} \times \vec{B}$ (right) for two vectors $\vec{A}$ and $\vec{B}$. The angle between the two vectors is ψ.

For now, though, it's just important to realize that if we write some vector as

$$\begin{aligned} \vec{A} &= A_x\hat{i} + A_y\hat{j} + A_z\hat{k} \\ \text{or} \quad &= A_{x'}\hat{i}' + A_{y'}\hat{j}' + A_{z'}\hat{k}' \end{aligned}$$

then it is still the same vector, even though the magnitudes of the components A_x, A_y, and A_z do not need to be the same in the "primed" coordinate system, that is $A_{x'}$, $A_{y'}$, and $A_{z'}$.

4.1.2 INNER PRODUCT AND CROSS PRODUCT

The *inner product*, also known as the *scalar product* or the *dot product*, of two vectors $\vec{A}$ and $\vec{B}$ is a geometric quantity. I will define it as the product of the magnitude of $\vec{A}$ times the magnitude of $\vec{B}$ times the cosine of the angle ψ between them. That is

$$\vec{A} \cdot \vec{B} = |\vec{A}||\vec{B}|\cos\psi \tag{4.4}$$

A simple interpretation of this formula is that the dot product is the projection of $\vec{A}$ in the direction of $\vec{B}$, times the magnitude of $\vec{B}$. See Figure 4.1. (Equivalently, we could say the dot product is the projection of $\vec{B}$ in the direction of $\vec{A}$, times the magnitude of $\vec{A}$.) This makes it very clear that the value of $\vec{A} \cdot \vec{B}$ has nothing to do with the coordinate system being used, including the possibility of a rotated set of Cartesian axes.

It is clear from (4.4) that the order of the vectors in the inner product does not matter. That is

$$\vec{A} \cdot \vec{B} = \vec{B} \cdot \vec{A}$$

An obvious byproduct of this definition is that the inner product of a vector with itself is the square of its magnitude, that is

$$\vec{A} \cdot \vec{A} = |\vec{A}|^2$$

If we apply this to unit vectors, then it is clear that the inner product of any unit vector with itself is unity. For example

$$\hat{\imath}\cdot\hat{\imath} = 1 = \hat{\jmath}\cdot\hat{\jmath} = \hat{k}\cdot\hat{k} = \hat{r}\cdot\hat{r} = \hat{\phi}\cdot\hat{\phi} = \hat{\theta}\cdot\hat{\theta}$$

Another obvious by product is that if two vectors $\vec{A}$ and $\vec{B}$ are perpendicular to each other, that is $\psi = 90^\circ$, then $\vec{A}\cdot\vec{B} = 0$. In this case, we say that $\vec{A}$ and $\vec{B}$ are *orthogonal* to each other.

Now the unit vectors are all at 90° with respect to each other, in a given coordinate system. That is, they are orthogonal to each other. Mathematically,

$$\begin{aligned} 0 &= \hat{\imath}\cdot\hat{\jmath} = \hat{\jmath}\cdot\hat{k} = \hat{k}\cdot\hat{\imath} && \text{Cartesian} \\ 0 &= \hat{r}\cdot\hat{\phi} = \hat{\phi}\cdot\hat{k} = \hat{k}\cdot\hat{r} && \text{Cylindrical} \\ 0 &= \hat{r}\cdot\hat{\theta} = \hat{\theta}\cdot\hat{\phi} = \hat{\phi}\cdot\hat{r} && \text{Spherical} \end{aligned}$$

(Cylindrical and spherical coordinates are discussed in Section 4.1.3.) This allows us to write the dot product using coordinates. If we write

$$\begin{aligned} \vec{A} &= A_x\hat{\imath} + A_y\hat{\jmath} + A_z\hat{k} \\ \vec{B} &= B_x\hat{\imath} + B_y\hat{\jmath} + B_z\hat{k} \end{aligned}$$

and multiply out $\vec{A}\cdot\vec{B}$, then we get

$$\begin{aligned} \vec{A}\cdot\vec{B} &= A_xB_x\hat{\imath}\cdot\hat{\imath} + A_xB_y\hat{\imath}\cdot\hat{\jmath} + A_xB_z\hat{\imath}\cdot\hat{k} \\ &\quad + A_yB_x\hat{\jmath}\cdot\hat{\imath} + A_yB_y\hat{\jmath}\cdot\hat{\jmath} + A_yB_z\hat{\jmath}\cdot\hat{k} \\ &\quad + A_zB_x\hat{k}\cdot\hat{\imath} + A_zB_y\hat{k}\cdot\hat{\jmath} + A_zB_z\hat{k}\cdot\hat{k} \\ &= A_xB_x + A_yB_y + A_zB_z \end{aligned} \tag{4.5}$$

It is rather remarkable that a rotation of the axes, as described in Section 4.1.1, has to give the same value in terms of the "primed" components. In fact, we formulate rotations mathematically by insisting that the dot product be *invariant* under a rotation transformation.

Another geometric vector product is the *cross product* $\vec{A}\times\vec{B}$, also depicted in Figure 4.1. Unlike the dot product, which is a (real) number, the cross product is itself another vector. The magnitude of $\vec{C} = \vec{A}\times\vec{B}$ is the area of the parallelogram formed by $\vec{A}$ and $\vec{B}$. Obviously, this means that the cross product of a vector with itself is zero, that is $\vec{A}\times\vec{A} = 0$.

The direction of the cross product usually makes use of something called the "right hand rule" which has had physics students playing with their fingers for decades. I will be more precise about this shortly when we discuss cross product of unit vectors, but there is an easy way to visualize the direction of the cross product from Figure 4.1. Imagine that you are "turning" the direction of $\vec{A}$ into the direction $\vec{B}$, about an axis that is perpendicular to both of them.

Now imagine that you turning a typical screw in that direction. The direction that the screw advances is the direction of $\vec{C} = \vec{A}\times\vec{B}$. (The "typical" screw is right handed.) Of course, this means that $\vec{B}\times\vec{A} = -\vec{C}$. In other words

$$\vec{A}\times\vec{B} = -\vec{B}\times\vec{A}$$

Now consider the Cartesian unit vectors $\hat{\imath}$, $\hat{\jmath}$, and $\hat{k}$. They are orthogonal, so the magnitude of their cross products is just unity. How we define the direction of their cross products, though, will set the "handedness" of the coordinate system. The convention that everyone sticks with, defining a "right handed" coordinate system, is the following:

$$\hat{\imath}\times\hat{\jmath} = \hat{k} \tag{4.6a}$$
$$\hat{k}\times\hat{\imath} = \hat{\jmath} \tag{4.6b}$$
$$\hat{\jmath}\times\hat{k} = \hat{\imath} \tag{4.6c}$$

Notice how the vectors "cyclically rotate" through the three equations. If we reverse any one (or all three) of these definitions, the coordinate system would be "left handed."

The cross products of the unit vectors let us write out the cross product in terms of components, similar to the way we did it for the dot product. We have

$$\begin{aligned}\vec{A}\times\vec{B} &= A_xB_x\hat{\imath}\times\hat{\imath}+A_xB_y\hat{\imath}\times\hat{\jmath}+A_xB_z\hat{\imath}\times\hat{k} \\ &\quad +A_yB_x\hat{\jmath}\times\hat{\imath}+A_yB_y\hat{\jmath}\times\hat{\jmath}+A_yB_z\hat{\jmath}\times\hat{k} \\ &\quad +A_zB_x\hat{k}\times\hat{\imath}+A_zB_y\hat{k}\times\hat{\jmath}+A_zB_z\hat{k}\times\hat{k} \\ &= (A_yB_z-A_zB_y)\hat{\imath}+(A_zB_x-A_xB_z)\hat{\jmath}+(A_xB_y-A_yB_x)\hat{k}\end{aligned} \tag{4.7}$$

4.1.2.1 Calculations with components using δ_{ij} and ε_{ijk}

Physics problems will often make use of dot products and cross products of both vector fields and vector derivatives (Section 4.2). We want to be efficient about these kinds of calculations, and there are some good tools for this. The first thing we want to do is associate an index $i = 1,2,3$ with components x,y,z. That is, we can rewrite the dot product from (4.5) as

$$\vec{A}\cdot\vec{B} = \sum_{i=1}^{3} A_iB_i$$

You can see that this is going to write a lot of summation signs, so we will implement the *summation convention* which says that if an index is repeated in an expression, it is implied that we need to sum over that index, setting it equal to 1, 2, and 3, and adding up the terms. Therefore, we write

$$\vec{A}\cdot\vec{B} = A_iB_i$$

where the sum over i is implied. We will never allow ourselves to be in a situation where we need to sum over an index that appears more than twice. In fact, no expression should ever be written down where any index appears three times or more.

For the unit vectors we can write $\hat{e}_1 \equiv \hat{\imath}$, $\hat{e}_2 \equiv \hat{\jmath}$, $\hat{e}_3 \equiv \hat{k}$, so

$$\vec{A} = A_i \hat{e}_i$$

is a handy way to write a vector. A second vector might be $\vec{B} = B_j \hat{e}_j$, where a different summation index j is used instead of i. In this case the dot product would be written

$$\vec{A} \cdot \vec{B} = A_i \hat{e}_i \cdot B_j \hat{e}_j = A_i B_j \, \hat{e}_i \cdot \hat{e}_j$$

where now this represents nine terms, summing over both i and j. Of course $\hat{e}_i \cdot \hat{e}_j$ equals unity if $i = j$ and is zero otherwise, so the nine terms collapse to the three terms represented by $A_i B_i$. There is a very convenient way to write this using the *Kronecker delta* δ_{ij}, defined simply as

$$\begin{aligned} \delta_{ij} &= 1 \qquad \text{for} \qquad i = j \\ &= 0 \qquad \text{for} \qquad i \neq j \end{aligned}$$

That is $\hat{e}_i \cdot \hat{e}_j = \delta_{ij}$ and we write

$$\vec{A} \cdot \vec{B} = A_i B_j \, \delta_{ij} = A_i B_i$$

The Kronecker delta effectively lets you "get rid of an index" in the implied sum.

There is a similar symbol that helps us work with cross products. I think it is more or less officially called the *Levi-Civita symbol for three dimensions*, but most people I know refer to it as the *totally antisymmetric symbol.* We write it as ε with three indices that I'll call i, j, and k (not to be confused with the names of the unit vectors!) The definition is

$$\begin{aligned} \varepsilon_{123} &= +1 \\ \text{and} \qquad \varepsilon_{ijk} &= -\varepsilon_{jik} = -\varepsilon_{ikj} - \varepsilon_{kji} \end{aligned}$$

In other words, $\varepsilon_{ijk} = 1$ if the i, j, k are in standard right-handed order, and flipping any two indices reverses the sign. Clearly, then, if any two indices are the same, then $\varepsilon_{ijk} = 0$. We can then write the cross product $\vec{A} \times \vec{B}$ as

$$\vec{A} \times \vec{B} = \varepsilon_{ijk} \hat{e}_i A_j B_k$$

a sum which, technically, has 9 terms, three of which are zero. We can also write

$$(\vec{A} \times \vec{B})_i = \varepsilon_{ijk} A_j B_k$$

which gives the ith component of $\vec{A} \times \vec{B}$.

A very useful theorem which connects the totally antisymmetric symbol with the Kronecker delta, written using our summation convention, is

$$\varepsilon_{ijk} \varepsilon_{imn} = \delta_{jm} \delta_{kn} - \delta_{jn} \delta_{km} \tag{4.8}$$

Note that the left side sums over i. I'm not going to bother trying to prove this, but if you want to write out some or all of the 81 equations represented here, that's up to you. Let's use this to prove something we generally call the "back-cab" rule, namely

$$\vec{A} \times (\vec{B} \times \vec{C}) = \vec{B}(\vec{A} \cdot \vec{C}) - \vec{C}(\vec{A} \cdot \vec{B})$$

If we write this out for the ith component of the triple cross product, we get

$$\begin{aligned}
\left[\vec{A} \times (\vec{B} \times \vec{C})\right]_i &= \varepsilon_{ijk} A_j (\vec{B} \times \vec{C})_k = \varepsilon_{ijk} A_j \varepsilon_{kmn} B_m C_n = \varepsilon_{kij} \varepsilon_{kmn} A_j B_m C_n \\
&= (\delta_{im}\delta_{jn} - \delta_{in}\delta_{jm}) A_j B_m C_n = A_j B_i C_j - A_j B_j C_i \\
&= B_i(\vec{A} \cdot \vec{C}) - C_i(\vec{A} \cdot \vec{B})
\end{aligned}$$

which is the "back-cab" rule for the ith component of $\vec{A} \times (\vec{B} \times \vec{C})$. Notice that I used a cyclic permutation (or, if you prefer, two index flips) to turn ε_{ijk} into ε_{kij} so that I could use (4.8), albeit with different index notation.

4.1.3 POLAR, CYLINDRICAL, AND SPHERICAL COORDINATES

Figure 1.4 shows how to locate a point in a plane using either Cartesian coordinates (x, y) or *plane polar* coordinates (r, ϕ). and how to locate a point in three-dimensional space either Cartesian coordinates (x, y, z) or *spherical polar* coordinates (r, θ, ϕ).

There is a third way to locate a point in three-dimensional space, where plane polar coordinates are used in the xy plane, but the z-coordinate is intact. These are called *cylindrical polar* coordinates. Extending plane polar coordinates into cylindrical coordinates, however, introduces a notational complication. Since r will measure the distance from the origin to a point in three dimensions, it does not make sense to use r as the distance measure in the (x, y) plane. Instead, we write ρ for the radial coordinate in the (x, y) plane.

Therefore, we write a vector $\vec{A}$ in three dimensions as

$$\vec{A} = A_x \hat{i} + A_y \hat{j} + A_z \hat{k} \quad \text{(Cartesian)} \tag{4.9a}$$

$$\vec{A} = A_r \hat{r} + A_\theta \hat{\theta} + A_\phi \hat{\phi} \quad \text{(Spherical)} \tag{4.9b}$$

$$\vec{A} = A_\rho \hat{\rho} + A_\phi \hat{\phi} + A_z \hat{k} \quad \text{(Cylindrical)} \tag{4.9c}$$

The unit vectors $\hat{r}$, $\hat{\theta}$, $\hat{\phi}$, and $\hat{\rho}$ have the same meaning as in the Cartesian case, *but these unit vectors have a direction that depends on the position of the point in space to which they refer.* This makes a huge difference if we ever need to calculate something like $\partial \vec{A}/\partial x$, since we need to consider the derivatives of the components *as well as the unit vectors*. Let's see how this works.

Let's start with plane polar coordinates. Figure 1.4 makes it clear that

$$x = r\cos\phi \tag{4.10a}$$

$$y = r\sin\phi \tag{4.10b}$$

Now imagine that we change the position $\vec{r}$ a tiny bit $d\vec{r} = dx\hat{\imath} + dy\,\hat{\jmath}$. This means that x and y change by the infinitesimal amounts

$$\begin{aligned} dx &= \cos\phi\, dr - r\sin\phi\, d\phi \\ dy &= \sin\phi\, dr + r\cos\phi\, d\phi \end{aligned}$$

To find the unit vectors $\hat{r}$ ad $\hat{\phi}$, we consider the change $d\vec{r}$ happening in those two specific directions. To see what happens in the r-direction, set $d\phi = 0$. This gives

$$d\vec{r} = \hat{\imath}\cos\phi\, dr + \hat{\jmath}\sin\phi\, dr = (\hat{\imath}\cos\phi + \hat{\jmath}\sin\phi)dr = \hat{r}\,dr$$

where $\hat{r} = \hat{\imath}\cos\phi + \hat{\jmath}\sin\phi$ because it is clear that dr is the magnitude of $d\vec{r}$ when the change only happens in the r-direction. If instead we move in the ϕ-direction and set $dr = 0$, we get

$$d\vec{r} = -\hat{\imath}r\sin\phi\, d\phi + \hat{\jmath}r\cos\phi\, d\phi = (-\hat{\imath}\sin\phi + \hat{\jmath}\cos\phi)rd\phi = \hat{\phi}\,rd\phi$$

where $\hat{\phi} = -\hat{\imath}\sin\phi + \hat{\jmath}\cos\phi$ because the distance along an arc at radius r through an angle $d\phi$ is $rd\phi$, and that is the change in the vector $\vec{r}$ when we move only in the ϕ-direction.

To summarize, then, the unit vectors for plane polar coordinates r and ϕ are

$$\hat{r} = \hat{\imath}\cos\phi + \hat{\jmath}\sin\phi \tag{4.11a}$$

$$\hat{\phi} = -\hat{\imath}\sin\phi + \hat{\jmath}\cos\phi \tag{4.11b}$$

The position vector is

$$\vec{r} = \hat{\imath}\,x + \hat{\jmath}y = r(\hat{\imath}\cos\phi + \hat{\jmath}\sin\phi) = r\hat{r} \tag{4.12}$$

and an infinitesimal change in the position vector is

$$d\vec{r} = \hat{\imath}dx + \hat{\jmath}dy = dr\hat{r} + rd\phi\,\hat{\phi} \tag{4.13}$$

Figure 4.2 compares the unit vector orientations for Cartesian and plane polar coordinates in two dimensions. For cylindrical coordinates in three dimensions, we simply include the unit vector $\hat{k}$ in the z-direction, and change the "r" to a "ρ." This is the same unit vector and coordinate as in the Cartesian system.

The situation is very similar for spherical polar coordinates. Figure 1.4 shows that the projection of the radial coordinate onto the xy plane is $r\sin\theta$. The transformation equations are therefore

$$x = r\sin\theta\cos\phi \tag{4.14a}$$

$$y = r\sin\theta\sin\phi \tag{4.14b}$$

$$z = r\cos\theta \tag{4.14c}$$

which leads to the spherical unit vectors in terms of the Cartesian unit vectors as

$$\hat{r} = \hat{\imath}\sin\theta\cos\phi + \hat{\jmath}\sin\theta\sin\phi + \hat{k}\cos\theta \tag{4.15a}$$

$$\hat{\theta} = \hat{\imath}\cos\theta\cos\phi + \hat{\jmath}\cos\theta\sin\phi - \hat{k}\sin\theta \tag{4.15b}$$

$$\hat{\phi} = -\hat{\imath}\sin\phi + \hat{\jmath}\cos\phi \tag{4.15c}$$

Plane polar coordinates and unit vectors can now be written directly from the spherical case for $\theta = \pi/2$. Figure 4.2 also depicts the spherical unit vectors.

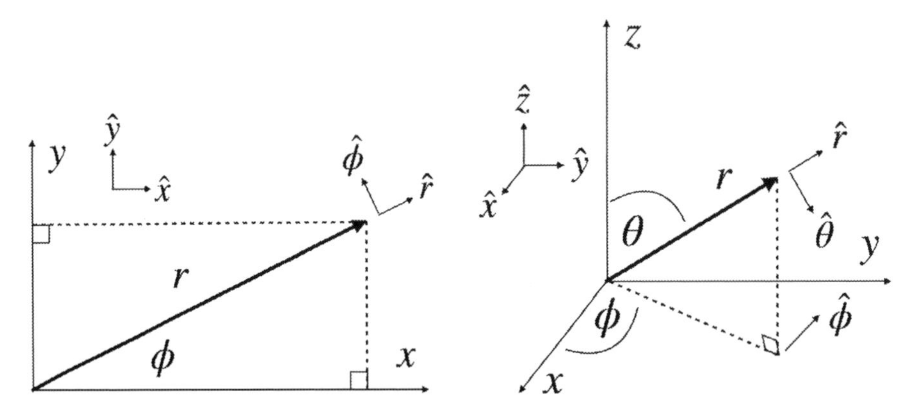

Figure 4.2 Cartesian and polar coordinate unit vectors in the plane (left) and in three dimensions (right). Recall Figure 1.4. It is important to note that while Cartesian unit vectors do not change their direction anywhere in the plane, the polar unit vectors indeed depend on r, θ, and ϕ. Note also that r in xy-plane is replaced by $r\sin\theta$ in three dimensions.

4.1.4 PATHS AND SURFACES

The position vector $\vec{r} = \hat{\imath}x + \hat{\jmath}y + \hat{k}z$ can be used to describe paths through space, or surfaces imbedded in space. For our purposes here, we'll assume the "space" is three dimensional. If we are working in two dimensions, then just leave off the z-coordinate. For paths and surfaces in higher dimensions, the generalization is relatively straightforward, but we won't pursue that here.

A path would be described by

$$\vec{r}(t) = \hat{\imath}x(t) + \hat{\jmath}y(t) + \hat{k}z(t) \tag{4.16}$$

where t is some parameter and $x(t)$, $y(t)$, and $z(t)$ are continuous functions of t. I use t because, in many problems in mechanics, for example, it stands for "time," but for now, we'll just assume it is some generic parameter. The point is that a position vector takes on different values as t changes, so changing t traces out a path in space.

A surface would be described by

$$f(x,y,z) = 0 \tag{4.17}$$

for some function $f(x,y,z)$. For example, $f(x,y,z) = x^2 + y^2 + z^2 - R^2$ describes s sphere of radius R that is centered at the origin. The point here is that the function $f(x,y,z)$ constrains the values of the coordinates so that, effectively, only two coordinates are needed to specify the position vector.

We'll make some us of paths and surfaces in this course, but you will make lots of use of these concepts in other physics courses. Right now, though, I want to focus on one specific kind of path, and one specific kind of surface.

4.1.4.1 Straight line paths

Your intuition tells you that you should be able to specify a straight line given its direction and one point through which it passes. Your intuition also tells you that $x(t)$, $y(t)$, and $z(t)$ should all vary with t the same way, so that they change proportionally to each other. We can combine these concepts by writing

$$x(t) = m_x t + x_0 \qquad y(t) = m_y t + y_0 \qquad z(t) = m_z t + x_0$$

where m_x, m_y, and m_z are just some constants. Clearly, this line passes through the point (x_0, y_0, z_0) when $t = 0$.

This immediately tells you, of course, that you can specify a straight line path by writing

$$\frac{x - x_0}{m_x} = \frac{y - y_0}{m_y} = \frac{z - z_0}{m_z}$$

which gets the job done, albeit a little sloppily. We can do a better job, though. The position vector (4.16) becomes

$$\vec{r}(t) = \vec{m}t + \vec{r}_0 \tag{4.18}$$

where $\vec{m} = \hat{\imath}m_x + \hat{\jmath}m_y + \hat{k}m_z$ is the direction of the line, and $\vec{r}_0 = \hat{\imath}x_0 + \hat{\jmath}y_0 + \hat{k}z_0$ is the position vector of a point through which the line passes. Sometimes it might be more convenient to use a unit vector $\hat{m}$ in the direction $\vec{m}$, and redefine any dimensioned quantities appropriately.

4.1.4.2 Planar surfaces

In order to specify a plane in three dimensions, we first need to identify a point $\vec{r}_0$ that is in the plane, and then write down an expression for the points $\vec{r}$ such that the vector from $\vec{r}_0$ to $\vec{r}$ lies in the plane. This is easy to do if we specify a vector $\hat{n}$ that is normal to the plane. See Figure 4.3. This gives the equation of the plane as

$$\hat{n} \cdot (\vec{r} - \vec{r}_0) = 0 \tag{4.19}$$

More convenient forms might multiply through by some constant so that the normal vector has a different physical meaning, for example the wave vector of a plane wave. A generic algebraic form for the equation of a plane might be written as

$$Ax + By + Cz = D$$

for constants A, B, C, and D.

4.1.5 VECTOR GENERALIZATIONS

When we get to Chapter 6 and Section 6.2, we will talk about the generalization of vectors to include *vector spaces*. Manipulating these generalized vectors will involve matrices and matrix operations, and will have very many applications to problems in the physical sciences.

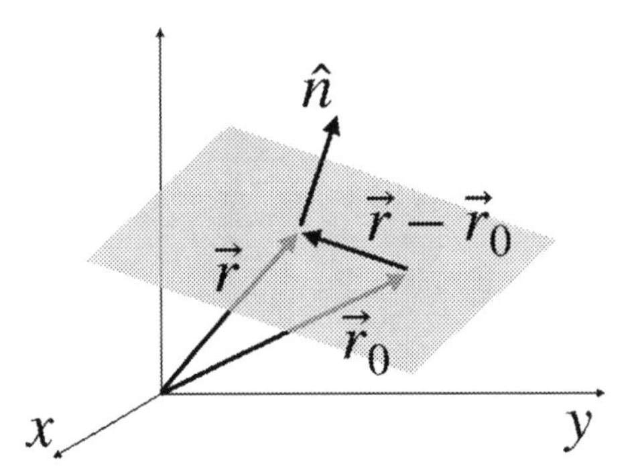

Figure 4.3 The locus of points $\vec{r}=\hat{\imath}x+\hat{\jmath}y+\hat{k}z$ form a plane in three dimensions that includes the point $\vec{r}_0$ if the collection of vectors $\vec{r}-\vec{r}_0$ are all perpendicular to some normal unit vector $\hat{n}$. That is, the equation of the plane is $\hat{n}\cdot(\vec{r}-\vec{r}_0)=0$.

Nevertheless, I want to mention here some generalizations that still maintain the ideas of vectors in physical space. Probably the first one that comes to mind is a concept you may encounter when studying Special Relativity, namely that "time is the fourth dimension." Indeed, the "three-vectors" we have been discussing here become "four-vectors" when including time as a dimension. We call this four-dimensional space *spacetime*. In this case, the vectors have four elements, with the time-like element having the index zero. A convention is that we use Latin indices like i, j, and k when we run over the three spatial dimensions, and Greek indices like μ, ν, and σ when we run over 0, 1, 2, and 3.

We can denote the four-vectors of spacetime as $\mathsf{A}=(A_0,A_1,A_2,A_3)=(A_0,\vec{A})$. The most important difference when moving from vectors in space to spacetime is in the dot product. In this case, the dot product must be invariant under Lorentz Transformation, which implies that the dot product between two four-vectors A and B becomes

$$\mathsf{A}\cdot\mathsf{B}=A_0B_0-A_1B_1-A_2B_2-A_3B_3=A_0B_0-\vec{A}\cdot\vec{B}$$

where the relative minus sign between the spacelike and timelike components has many physical implications.

A more general application of spacetime, which is necessary for understanding General Relativity and the theory of gravitation, comes from the need to recognize that there are in fact two different geometric classes of spatial vectors. These are called *contravariant vectors* and *covariant vectors*, aka "vectors" and "covectors."[1]

4.2 VECTOR OPERATORS

Now we turn our attention to *spatial differential vector operators*. These are constructs that take (partial) derivatives of functions of spatial position, i.e. *fields*. We

[1] If you want to learn a little more contravariant and covariant vectors, see "Answer to Question #55. Are there pictorial examples that distinguish covariant and contravariant vectors?," American Journal of Physics 65(1997)1037 by J. Napolitano and R. Lichtenstein, https://doi.org/10.1119/1.18743.

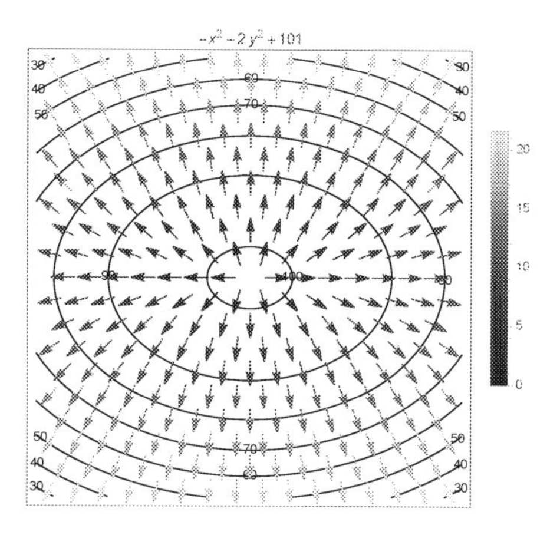

Figure 4.4 Visualizing the gradient of $f(\vec{r}) = f(x,y) = 101 - x^2 - 2y^2$, a scalar field in two spatial dimensions. The function $f(x,y)$ is drawn as a contour plot, with the contours labeled. The (negative of its) gradient $-\vec{\nabla} f = 2x\hat{i} + 4y\hat{j}$ is drawn with arrows at different points (x,y) that point in the direction of the gradient, and whose graytone indicates the magnitude $|\vec{\nabla} f|$ at that point. Notice that the magnitude of the gradient is larger in the regions where the contours are more steep. (I plot $-\vec{\nabla} f$ instead of $\vec{\nabla} f$ because it feels better to go down the hill instead of up it.)

will construct these operators first in Cartesian coordinates, but then also give them, with some derivations, in cylindrical and spherical coordinates.

Most of the material in this section should be reviewed from your calculus classes.

4.2.1 GRADIENT

Suppose you have a scalar field $f(\vec{r}) = f(x,y,z)$ (or $f(x,y)$ in two dimensions) over some region in space. You will typically need to know how fast that function changes, that is, you will need to know its derivative. However, the rate of change will depend on which direction in space you are moving. Somehow we need to come up with a "derivative" that respects the direction in space. In other words, we need a vector version of the derivative.

That vector version of the derivative is called the *gradient* of $f(\mathbf{r})$, written in Cartesian coordinates as

$$\vec{\nabla} f = \hat{i}\frac{\partial f}{\partial x} + \hat{j}\frac{\partial f}{\partial y} + \hat{k}\frac{\partial f}{\partial z}$$

This vector quantity, in principle different at any different point in space, will tell the direction in which the change of $f(\mathbf{r})$ is a maximum. Figure 4.4 is a visualization of the gradient for a particular two-dimensional scalar field. (Actually, it is the negative of the gradient that is plotted, only because it feels more natural to go down the hill instead of up it.) Notice how the magnitude of the gradient (given by the graytone of the arrow) is larger where the hill is steeper.

It is not hard to see that the gradient tells us the direction in which the change is $f(\vec{r})$ is greatest. Consider moving in some direction $\vec{s}$. The infinitesimal change in the distance along $\vec{s}$ is ds, so the derivative of $f(\vec{r})$ in this direction is

$$\begin{aligned}\frac{df}{ds} &= \frac{1}{ds}\left[\frac{\partial f}{\partial x}dx+\frac{\partial f}{\partial y}dy+\frac{\partial f}{\partial z}dz\right]=\vec{\nabla}f\cdot\left[\hat{i}\frac{dx}{ds}+\hat{j}\frac{dy}{ds}+\hat{k}\frac{dz}{ds}\right]=\vec{\nabla}f\cdot\hat{s}\\ \text{where}\quad \hat{s} &= \hat{i}\frac{dx}{ds}+\hat{j}\frac{dy}{ds}+\hat{k}\frac{dz}{ds}\end{aligned}$$

is the unit vector in the s-direction, since $ds^2 = dx^2+dy^2+dz^2$. Of course, $\vec{\nabla}f\cdot\hat{s} = |\vec{\nabla}f|\cos\psi$ where ψ is the angle between the gradient and $\vec{s}$. Therefore, the direction of greatest change is when $\psi = 0$, that is, in the direction of the gradient.

We will find it handier to think of $\vec{\nabla}$ as a *vector operator* which we write as

$$\vec{\nabla} = \hat{i}\frac{\partial}{\partial x}+\hat{j}\frac{\partial}{\partial y}+\hat{k}\frac{\partial}{\partial z} \qquad \text{Cartesian coordinates} \tag{4.20}$$

In other words, $\vec{\nabla}f$ is the gradient operator "acting on" the field $f(\vec{r})$. Similarly, we can think of $\hat{i}\cdot\vec{\nabla} = \partial/\partial x$ as a differential operator and so forth.

It is straightforward to write down the gradient operator in different coordinate systems. For example, if $f(\vec{r})$ is expressed in cylindrical coordinates (ρ,ϕ,z), then

$$\vec{\nabla}f = \hat{i}\frac{\partial f(\rho,\phi,z)}{\partial x}+\hat{j}\frac{\partial f(\rho,\phi,z)}{\partial y}+\hat{k}\frac{\partial f(\rho,\phi,z)}{\partial z}$$

and we use the chain rule to express the derivatives in terms of (ρ,ϕ,z). We have

$$\frac{\partial f}{\partial x} = \frac{\partial f}{\partial \rho}\frac{\partial \rho}{\partial x}+\frac{\partial f}{\partial \phi}\frac{\partial \phi}{\partial x}+\frac{\partial f}{\partial z}\frac{\partial z}{\partial x}$$

Inverting (4.10), but writing ρ instead of r, gives

$$\rho = \left(x^2+y^2\right)^{1/2} \qquad \text{so} \qquad \frac{\partial \rho}{\partial x} = \frac{x}{(x^2+y^2)^{1/2}} = \frac{x}{\rho} = \cos\phi$$

and

$$\tan\phi = \frac{y}{x} \qquad \text{so} \qquad \frac{1}{\cos^2\phi}\frac{\partial \phi}{\partial x} = -\frac{y}{x^2} = \frac{-\rho\sin\phi}{\rho^2\cos^2\phi} \qquad \text{and} \qquad \frac{\partial \phi}{\partial x} = -\frac{1}{\rho}\sin\phi$$

Of course, $\partial z/\partial x = 0$ so the first term of $\vec{\nabla}f$ in cylindrical coordinates is

$$\hat{i}\frac{\partial f(\rho,\phi,z)}{\partial x} = \hat{i}\left[\frac{\partial f}{\partial \rho}\cos\phi+\frac{1}{\rho}\frac{\partial f}{\partial \phi}(-\sin\phi)+0\right] = \hat{i}\left[\cos\phi\frac{\partial f}{\partial \rho}-\sin\phi\frac{1}{\rho}\frac{\partial f}{\partial \phi}\right]$$

(Note the $1/\rho$ factor in front of the ϕ derivatives, which makes the expression dimensionally correct.) Similarly, the second term is

$$\hat{j}\frac{\partial f(\rho,\phi,z)}{\partial y} = \hat{j}\left[\sin\phi\frac{\partial f}{\partial \rho}+\cos\phi\frac{1}{\rho}\frac{\partial f}{\partial \phi}\right]$$

The third term is just $\hat{k}\partial f/\partial z$, or simply missing if we are working in plane polar coordinates. Putting this all together, the gradient becomes

$$\begin{aligned}\vec{\nabla} f &= \hat{i}\left[\cos\phi\frac{\partial f}{\partial\rho} - \sin\phi\frac{1}{\rho}\frac{\partial f}{\partial\phi}\right] + \hat{j}\left[\sin\phi\frac{\partial f}{\partial\rho} + \cos\phi\frac{1}{\rho}\frac{\partial f}{\partial\phi}\right] + \hat{k}\frac{\partial f}{\partial z} \\ &= (\hat{i}\cos\phi + \hat{j}\sin\phi)\frac{\partial f}{\partial\rho} + (-\hat{i}\sin\phi + \hat{j}\cos\phi)\frac{1}{\rho}\frac{\partial f}{\partial\phi} + \hat{k}\frac{\partial f}{\partial z} \\ &= \hat{r}\frac{\partial f}{\partial\rho} + \hat{\phi}\frac{1}{\rho}\frac{\partial f}{\partial\phi} + \hat{k}\frac{\partial f}{\partial z}\end{aligned}$$

where we made use of the expressions (4.11) for the unit vectors in plane polar coordinates. This means that the gradient operator in cylindrical coordinates is

$$\vec{\nabla} = \hat{r}\frac{\partial}{\partial r} + \hat{\phi}\frac{1}{\rho}\frac{\partial}{\partial\phi} + \hat{k}\frac{\partial}{\partial z} \qquad \text{Cylindrical coordinates} \qquad (4.21)$$

It's nice to see this carried out from the fundamentals, but there is in fact an easier way to get this result. In plane polar coordinates, the infinitesimal change in the position vector is given by (4.13). If you think of the gradient as the directional derivative, simple inspection shows that if you move in the radial direction only, the derivative is $\partial/\partial\rho$. On the other hand, if you move in the axial direction only, the derivative is $(1/\rho)\partial/\partial\phi$. Including the z-coordinate, then, the gradient operator is clearly (4.21).

For spherical coordinates, you carry through exactly the same way, but using (4.14) instead of (4.11). You find

$$\vec{\nabla} = \hat{r}\frac{\partial}{\partial r} + \hat{\theta}\frac{1}{r}\frac{\partial}{\partial\theta} + \hat{\phi}\frac{1}{r\sin\theta}\frac{\partial}{\partial\phi} \qquad \text{Spherical coordinates} \qquad (4.22)$$

I'll leave deriving this as a homework problem.

4.2.2 DIVERGENCE

We can define other operators based on the gradient operator. Given a vector field

$$\vec{V}(\vec{r}) = \vec{V}(x,y,z) = \hat{i}V_x(x,y,z) + \hat{j}V_y(x,y,z) + \hat{k}V_z(x,y,z) \qquad (4.23)$$

we can use the gradient operator to measure how much the field "diverges" in a region of space. I'll make clearer what I mean by that in Section 4.3.2, but for now just think of it measuring how much of $\vec{V}(\vec{r})$ has a "flow" that emerges from a region of space.

We define the *divergence* of a vector field $\vec{V}(\vec{r}) = \vec{V}(x,y,z)$ as

$$\vec{\nabla}\cdot\vec{V} = \frac{\partial V_x}{\partial x} + \frac{\partial V_y}{\partial y} + \frac{\partial V_z}{\partial z} \qquad (4.24a)$$

where we are literally taking the dot product of the vector operator $\vec{\nabla}$ with the vector field $\vec{V}$. It is very important to note that when we did this, we treated the unit vectors

$\hat{i}$, $\hat{j}$, and $\hat{k}$ as constants. Indeed, they do not depend on position. However, this will not be the case when we consider cylindrical and spherical coordinates.

I'm not going to bother deriving the form of the gradient in cylindrical or spherical coordinates. (I also don't think they make good homework problems.) They are tedious calculations, and not, in my opinion, particularly instructive, although it is a good practice in partial derivatives and the chain rule to carry through the calculation. Instead, I will just state the results, which you can find many places online or in textbooks. The results are

$$\vec{\nabla}\cdot\vec{V} = \frac{1}{\rho}\frac{\partial}{\partial\rho}(\rho V_\rho) + \frac{1}{\rho}\frac{\partial V_\phi}{\partial\phi} + \frac{\partial V_z}{\partial z} \tag{4.24b}$$

in cylindrical polar coordinates, and

$$\vec{\nabla}\cdot\vec{V} = \frac{1}{r^2}\frac{\partial}{\partial r}(r^2 V_r) + \frac{1}{r\sin\theta}\frac{\partial}{\partial\theta}(\sin\theta\, V_\theta) + \frac{1}{r\sin\theta}\frac{\partial V_\phi}{\partial\phi} \tag{4.24c}$$

in spherical polar coordinates.

4.2.3 CURL

Applying the cross product of the gradient operator to a vector field is called the *curl*, for reasons that will become apparent in Section 4.3.1. We have

$$\vec{\nabla}\times\vec{V} = \hat{i}\left[\frac{\partial V_z}{\partial y} - \frac{\partial V_y}{\partial z}\right] + \hat{j}\left[\frac{\partial V_x}{\partial z} - \frac{\partial V_z}{\partial x}\right] + \hat{k}\left[\frac{\partial V_y}{\partial x} - \frac{\partial V_x}{\partial y}\right] \tag{4.25a}$$

$$\begin{aligned}\vec{\nabla}\times\vec{V} &= \hat{r}\left[\frac{1}{\rho}\frac{\partial V_z}{\partial\phi} - \frac{\partial V_\phi}{\partial z}\right] + \hat{\phi}\left[\frac{\partial V_\rho}{\partial z} - \frac{\partial V_z}{\partial\rho}\right] \\ &\quad + \hat{k}\frac{1}{\rho}\left[\frac{\partial}{\partial\rho}(\rho V_\phi) - \frac{\partial V_\rho}{\partial\phi}\right]\end{aligned} \tag{4.25b}$$

$$\begin{aligned}\vec{\nabla}\times\vec{V} &= \hat{r}\frac{1}{r\sin\theta}\left[\frac{\partial}{\partial\theta}V_\phi\sin\theta - \frac{\partial V_\theta}{\partial\phi}\right] + \hat{\theta}\frac{1}{r}\left[\frac{1}{\sin\theta}\frac{\partial V_r}{\partial\phi} - \frac{\partial}{\partial r}rV_\phi\right] \\ &\quad + \hat{\phi}\frac{1}{r}\left[\frac{\partial}{\partial r}rV_\theta - \frac{\partial V_r}{\partial\theta}\right]\end{aligned} \tag{4.25c}$$

in Cartesian, cylindrical polar, and spherical polar coordinates, respectively.

Although it may be more or less obvious from general ideas about vector cross products and dot products, it is worth stating outright that (1) the curl of any gradient is zero and (2) the divergence of any curl is zero. That is

$$\vec{\nabla}\times\vec{\nabla}f(\vec{r}) = 0$$

for any scalar field $f(\vec{r})$, and

$$\vec{\nabla}\cdot\left[\vec{\nabla}\times\vec{V}(\vec{r})\right] = 0$$

for any vector field $\vec{V}(\vec{r})$. You will see in your electromagnetism classes that this is the basis for defining the scalar and vector potentials. See also the example in Section 4.5.1.2.

4.2.4 THE LAPLACIAN

You will encounter many calculations in physics where you want to take the divergence of some vector field, which itself is the gradient of some scalar field.[2] This leads to the (scalar) operator $\vec{\nabla}^2 = \vec{\nabla} \cdot \vec{\nabla}$ known as the *Laplacian*. Its form in Cartesian coordinates is obvious, namely

$$\vec{\nabla}^2 = \frac{\partial^2}{\partial x^2} + \frac{\partial^2}{\partial y^2} + \frac{\partial^2}{\partial z^2} \tag{4.26a}$$

The form of the Laplacian in cylindrical or spherical coordinates is less obvious, but straightforward to derive. As always, you need to remember that that unit vectors in this case need to be differentiated because they depend on position. One finds

$$\vec{\nabla}^2 = \frac{1}{\rho}\frac{\partial}{\partial \rho}\left(\rho\frac{\partial}{\partial \rho}\right) + \frac{1}{\rho^2}\frac{\partial^2}{\partial \phi^2} + \frac{\partial^2}{\partial z^2} \tag{4.26b}$$

in cylindrical polar coordinates, and

$$\vec{\nabla}^2 = \frac{1}{r^2}\frac{\partial}{\partial r}\left(r^2\frac{\partial}{\partial r}\right) + \frac{1}{r^2 \sin\theta}\frac{\partial}{\partial \theta}\left(\sin\theta\frac{\partial}{\partial \theta}\right) + \frac{1}{r^2 \sin^2\theta}\frac{\partial^2}{\partial \phi^2} \tag{4.26c}$$

in spherical polar coordinates. You will find later that the radial part of (4.26b) gives rise to Bessel's Equation (3.23), the radial part of (4.26c) leads to spherical Bessel functions, and the θ equation from (4.26c) becomes Legendre's Equation (3.33).

It should be clear that you can apply the Laplacian operator to either a scalar field or a vector field. In case of the latter, you just need to apply it to each component, which can be tricky if you are applying the operator in cylindrical or spherical coordinates to a vector field in cylindrical or spherical coordinates.

4.3 SURFACE THEOREMS

Probably the most important physical applications of vector calculus have to do with a class of properties that I'll call *surface theorems*. We will discuss two of these, in particular, namely Stokes' Theorem and Gauss' Theorem. (I think mathematicians probably refer to Gauss' Theorem as the Divergence Theorem.)

The idea behind the surface theorems is that whatever is going on inside some closed region can be inferred from what is happening on the surface of that region. Stokes' Theorem applies to regions enclosed by a loop in space, and Gauss' Theorem applies to volumes and the surfaces that enclose them.

This section will first state and then prove Stokes' Theorem first, then Gauss' Theorem. These two theorems are intimately connected to the curl and divergence, respectively, of a vector field. In fact, a better way to present the curl and divergence would be to see how they arise naturally from these theorems. This is the approach taken in a popular text from many years ago, called "Div, Grad, Curl, and All That: An Informal Text on Vector Calculus," by H.M. Schey. It's a nice book, and it shouldn't be too hard for you to locate a copy if you would like to look it over.

[2]Probably the first example of this is taking the divergence of the electric field, written as the (negative) gradient of the electric potential.

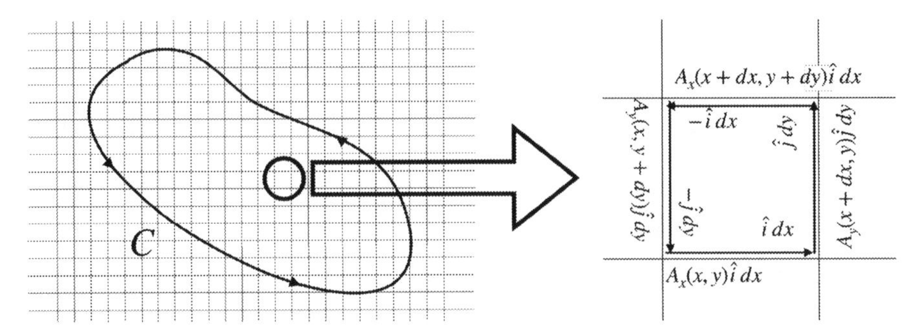

Figure 4.5 Figure used in the proof of Stokes' Theorem. An arbitrary curve C bounds some surface S, which in this example lies in the xy plane. The plane is divided up into a bunch of infinitesimal rectangles with dimension $dx \times dy$. The dot product $\vec{A} \cdot d\vec{\ell}$ is evaluated first along the sides of a tiny rectangle with lower left corner at the point (x, y). This is then generalized to include the entire curve C.

4.3.1 STOKES' THEOREM

Stokes' Theorem relates the integral of some vector quantity projected onto a closed loop to the curl of that vector field integrated over the surface enclosed by the loop. We state this theorem mathematically as

$$\oint_C \vec{A} \cdot d\vec{\ell} = \int_S (\vec{\nabla} \times \vec{A}) \cdot d\vec{S} \tag{4.27}$$

where C is the closed loop and S is the surface that it encloses. The shape of the surface doesn't matter, just so long as its edges lie along C. The integrand $\vec{A} \cdot d\vec{\ell}$ means the dot product of the vector field with a tiny line segment pointing along the line, and the little circle on the integral sign just means the line forms a closed loop. On the right hand integral, the vector $d\vec{S}$ is an infinitesimal whose magnitude dS is the size of the area element of S, and the direction is normal to the surface. The sign of the normal is determined by an agreement of using the right hand rule based on an assigned direction for following the loop.

We can prove Stokes' Theorem by first taking the surface S and carving it up into a bunch of tiny rectangles. We'll assume that S lies in the xy plane, and argue later that we can generalize to any curve. Each tiny rectangle has width dx and height dy. We'll then show that Stokes' Theorem holds for each tiny rectangle by itself.

Now if I put two tiny rectangles next to each other, then the line integral on the left side of (4.27) cancels along the adjacent edge, and the curve is over the pair of rectangles. Keep going, and eventually you'll cover the entire surface S, and the only curve that's left is C. At this point, we can say that we've proved Stokes' Theorem.

Figure 4.5 shows how this works. The sum of $\vec{A} \cdot d\vec{\ell}$ over the tiny square's sides is

$$\begin{aligned}\sum \vec{A} \cdot d\vec{\ell} &= A_x(x,y)dx + A_y(x+dx,y)dy - A_x(x+dx,y+dy)dx - A_y(x,y+dy)dy \\ &= [A_x(x,y) - A_x(x+dx,y+dy)]dx + [A_y(x+dx,y) - A_y(x,y+dy)]dy\end{aligned}$$

where we note that, for example, $\vec{A}\cdot d\vec{\ell} = A_x dx$ when $d\vec{\ell}$ is in the x-direction, and so forth. Care is taken to evaluate the components of the vector function consistently at the corner of the rectangle where piece of the integral starts. Now factoring out $dxdy$ and doing some rearranging, we get

$$\sum \vec{A}\cdot d\vec{\ell} = \left[\frac{A_y(x+dx,y) - A_y(x,y+dy)}{dx} - \frac{A_x(x+dx,y+dy) - A_x(x,y)}{dy}\right] dxdy$$

As the infinitesimals approach zero, the two expressions in the brackets become partial derivatives. The result is

$$\sum \vec{A}\cdot d\vec{\ell} = \left[\frac{\partial A_y}{\partial x} - \frac{\partial A_x}{\partial y}\right] dxdy = (\vec{\nabla}\times\vec{A})_z\, dxdy$$

where, finally, we recognize that combination of partial derivatives as the z-component of the curl. Since $dxdy = dS$ and, in this case, the normal to the surface is $\hat{z}$, we have, for the tiny rectangle,

$$\sum \vec{A}\cdot d\vec{\ell} = (\vec{\nabla}\times\vec{A})\cdot d\vec{S}$$

Adding up all the little rectangles, and recognizing that their edges cancel in the line integral leaving only the curve C, we get (4.27).

I've done this by assuming the surface is in a plane, dividing up into little parallel rectangles. However, I'll argue that the shape of the surface doesn't matter until after we've done Gauss' Theorem.

Let's illustrate this with a simple example. Take a vector field $\vec{A} = -y\hat{\imath} + x\hat{\jmath}$, which clearly has a "curl" to it if you plot it. In fact

$$\vec{\nabla}\times A = \left(\frac{\partial A_y}{\partial x} - \frac{\partial A_x}{\partial y}\right)\hat{k} = (1+1)\hat{k} = 2\hat{k}$$

Take the curve C to be the square in the xy plane with side length ℓ and the lower left corner at the origin. The surface S has area ℓ^2 so the right hand side of (4.27) is

$$\int (\vec{\nabla}\times\vec{A})\cdot d\vec{S} = 2\int dS = 2\ell^2$$

Now let's evaluate the line integral. Going counter clockwise from the lower left corner,

$$\begin{aligned}\oint \vec{A}\cdot d\vec{\ell} &= \int_{x=0}^{\ell} (-y)|_{y=0}\, dx + \int_{y=0}^{\ell} (x)|_{x=\ell}\, dy \\ &\quad + \int_{x=\ell}^{0} (-y)|_{y=\ell}\, dx + \int_{y=\ell}^{0} (x)|_{x=0}\, dy \\ &= 0 + \ell\cdot\ell - \ell\cdot(-\ell) = 2\ell^2\end{aligned}$$

as predicted by the theorem.

Note that for this field, with a constant curl, it shouldn't matter where we put the square! We'll evaluate it for the example given, but you should try putting it somewhere else, perhaps centered on the origin, to confirm that you get the same answer.

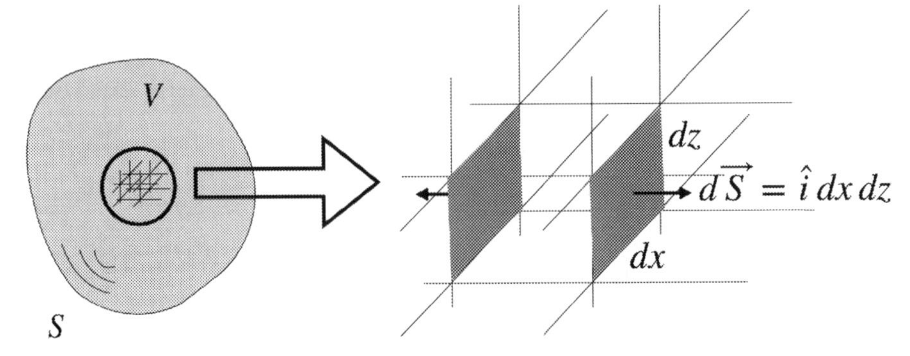

Figure 4.6 Figure used in the proof of Gauss' Theorem. An arbitrary three-dimensional volume V is enclosed by a surface S. The volume is divided up into little bricks with dimensions $dx \times dy \times dz$. For each little brick, you add up the six values of $\vec{A} \cdot d\vec{S}$. Then when you add up all of the little bricks to get the full volume, the surfaces of adjacent bricks cancel with each other, leaving only the surface S.

4.3.2 GAUSS' THEOREM

Gauss' Theorem (also known as the *Divergence Theorem*) relates the divergence of a vector field in some volume V to the surface integral of the normal component of the vector field over the surface S that encloses the volume. That is,

$$\oint_S \vec{A} \cdot d\vec{S} = \int_V \vec{\nabla} \cdot \vec{A}\, dV \tag{4.28}$$

When $\vec{A}$ is physically describing the "flow" of some quantity, in which case $\vec{A} \cdot d\vec{S}$ is the "flux" through the surface $d\vec{S}$, we will find this to be an extremely useful theorem.

The proof of Gauss' Theorem is very similar to how we proved Stokes' Theorem. See Figure 4.6. In this case, you chop up the volume V into a bunch of tiny bricks, with dimensions dx, dy, and dz. For one of these tiny bricks, the surface integral is the sum

$$\begin{aligned}
\sum \vec{A} \cdot d\vec{S} &= A_x(x+dx,y,z)(+dydz) + A_x(x,y,z)(-dydz) \\
&+ A_y(x,y+dy,z)(+dxdz) + A_y(x,y,z)(-dxdz) \\
&+ A_z(x,y,z+dz)(+dxdy) + A_z(x,y,z)(-dxdy)
\end{aligned}$$

where I've paired the terms by the "front" and "back" of each of the three directions. Factoring out $dxdydz = dV$, this expression becomes

$$\begin{aligned}
\sum \vec{A} \cdot d\vec{S} &= \left[\frac{A_x(x+dx,y,z) - A_x(x,y,z)}{dx} + \frac{A_y(x,y+dy,z) - A_y(x,y,z)}{dy} \right. \\
&\quad \left. + \frac{A_z(x,y,z+dz) - A_z(x,y,z)}{dy} \right] dxdydz \\
&= \left[\frac{\partial A_x}{\partial x} + \frac{\partial A_y}{\partial y} + \frac{\partial A_z}{\partial z} \right] dxdydz = \vec{\nabla} \cdot \vec{A}\, dV
\end{aligned}$$

When we add up all of the little bricks, the right side just becomes the volume integral. On the left side, adjacent sides of bricks cancel $\vec{A}\cdot d\vec{S}$ because the vector $d\vec{S}$ has the same magnitude but opposite direction, and all that is left from the sum is the surface S that encloses the volume V. This proves (4.28).

Let's do a simple example with the field $\vec{A} = x\hat{i} + y\hat{j} + z\hat{k}$. This field clearly "diverges," as a vector plot with any application will tell you. The divergence of this field is $\vec{\nabla}\cdot\vec{A} = 3$, that is, uniform everywhere, so the volume integral on the right side of (4.28) is just $3V$. A simple shape is a cube of side length ℓ that sits in the first octant with a corner at the origin. The flux $\vec{A}\cdot d\vec{S}$ is zero on the three faces that are in the three planes $x = 0$, $y = 0$, and $z = 0$ because the component perpendicular to the plane is zero there. The flux on the other three faces is just $\ell\cdot\ell^2 = \ell^3$, so the surface integral is $3\ell^3 = 3V$. It works.

4.3.2.1 Why the surface shape doesn't matter in Stokes' theorem

We gave a proof of (4.27) that assumed the surface was flat. Imagine there is now another surface that has the same curve C on the edge. These two surfaces form a closed volume, and the second surface has the opposite clockwise sense of the first. That is

$$\oint_{S_1+S_2} \vec{B}\cdot d\vec{S} = \int_{S_1} \vec{B}\cdot d\vec{S} - \int_{S_2} \vec{B}\cdot d\vec{S}$$

for some vector field $\vec{B}$. In the case of Stokes' Theorem, though, $\vec{B} = \vec{\nabla}\times\vec{A}$ for a different vector field $\vec{V}$. Now by Gauss' Theorem,

$$\oint_{S_1+S_2} \vec{B}\cdot d\vec{S} = \int_V \vec{\nabla}\cdot\vec{B} = \int_V \vec{\nabla}\cdot(\vec{\nabla}\times\vec{A}) = 0$$

since the divergence of any curl is zero. Therefore

$$\int_{S_1} \vec{B}\cdot d\vec{S} = \int_{S_2} \vec{B}\cdot d\vec{S}$$

so it doesn't matter which surface I use to be enclosed by C in Stokes' Theorem.

4.3.3 THE CONTINUITY EQUATION

The notion of a *conservation law* is fundamental in physics. We say that quantities like charge, energy, and momentum, for example, are "conserved." That is, they do not change with time. How can we write this down, mathematically, for quantities that can exist spread out over space?

We will explore the answer to this question by considering three-dimensional space. You can work out the problem for two-dimensional space, in which case the answer will involve the curl. That might make a nice homework problem.

Imagine that we have some quantity Q that is spread over some volume V of space. We would write Q in terms of some scalar field $\rho(\vec{r},t)$, namely the density of the material that makes up Q. The density has dimensions of $[Q]L^{-3}$, and it is possible that the density at any point can change with time t.

If Q is a conserved quantity, though, the only way it can change is if some of it flows into, our out of, the volume V. It is not possible to create or destroy the material that makes up Q if it is conserved. That is, there are no "sources" or "sinks" for this material inside V.

Define a vector field $\vec{j}(\vec{r},t)$ that represents the "flow" of this material. We'll write that the "flux" of this material through some area element $d\vec{S}$ is just $\vec{j}(\vec{r},t)\cdot d\vec{S}$. This makes sense because you recall that the direction of $d\vec{S}$ is perpendicular to the surface, so if the flow $\vec{j}(\vec{r},t)$ is perpendicular to the surface, then it is maximized, whereas it is zero if $\vec{j}(\vec{r},t)$ is parallel to the surface.

The dimensions of $\vec{j}(\vec{r},t)$ are $[Q]L^{-2}T^{-1}$ so the dimensions of the flux are $[Q]T^{-1}$. Sometimes we refer to $\vec{j}(\vec{r},t)$ as the "flux density," or "current density" particularly if Q represents electric charge.

Now consider the surface S that encloses the volume V. We know that the only way to change Q is for there to be some flux through S. Let's agree that a positive value for the flux is when the flow goes from the inside to the outside. This means that $\vec{j}(\vec{r},t)\cdot d\vec{S}$ is positive when $d\vec{S}$ is defined according to our normal convention that it points away from the inside of S. Integrating over the surface tells us how to write down the change of Q with time, namely

$$\frac{dQ}{dt} = -\oint_S \vec{j}(\vec{r},t)\cdot d\vec{S}$$

The minus sign tells us that if the net flux through the closed surface S is positive, then Q decreases, which is correct.

If we write Q in terms of the density $\rho(\vec{r},t)$ by integrating over the volume V, and we apply Gauss' Theorem to the surface integral, this equation becomes

$$\frac{d}{dt}\int_V \rho(\vec{r},t)\,dV = -\int_V \vec{\nabla}\cdot\vec{j}(\vec{r},t)\,dV$$

The time derivative is a total derivative because all that is left after doing the integral over position is time. I can bring the time derivative inside the integral, but then it becomes a partial derivative. This all leads us to

$$\int_V \left[\frac{\partial\rho(\vec{r},t)}{\partial t} + \vec{\nabla}\cdot\vec{j}(\vec{r},t)\right] dV = 0$$

Finally, we let the volume $V \to \Delta V$ be so small that we can neglect the change in the integrand over position and pull it out of the integral, leaving just the volume ΔV times the expression in square brackets, which itself now must be zero. Therefore

$$\frac{\partial\rho(\vec{r},t)}{\partial t} + \vec{\nabla}\cdot\vec{j}(\vec{r},t) = 0 \tag{4.29}$$

This differential equation is called the *Continuity Equation* and is the mathematical way to state that a quantity is conserved. It is used extensively in many fields of physics, including fluid dynamics, electrodynamics, biological systems, statistical mechanics, and even finance.

4.3.4 APPLICATION TO MAXWELL'S EQUATIONS

An important first application of the surface theorems is to turn Maxwell's Equations into their differential form. You likely learned the integral form of Maxwell's Equations in your introductory physics class, namely[3]

$$\oint_S \vec{E}\cdot d\vec{S} = 4\pi Q_{\text{enclosed}} = 4\pi\int_V \rho\, dV \qquad \text{Gauss' Law} \tag{4.30a}$$

$$\oint_S \vec{B}\cdot d\vec{S} = 0 \qquad \text{Gauss' Law for Magnetism} \tag{4.30b}$$

$$\oint_C \vec{E}\cdot d\vec{\ell} = -\frac{1}{c}\frac{d\Phi_B}{dt} = -\frac{1}{c}\frac{d}{dt}\int_S \vec{B}\cdot d\vec{S} \qquad \text{Faraday's Law} \tag{4.30c}$$

$$\begin{aligned}\oint_C \vec{B}\cdot d\vec{\ell} &= \frac{4\pi}{c}I_{\text{enclosed}} + \frac{1}{c}\frac{d\Phi_E}{dt} \\ &= \frac{4\pi}{c}\int_S \vec{j}\cdot d\vec{S} + \frac{1}{c}\frac{d}{dt}\int \vec{E}\cdot d\vec{S} \qquad \text{Ampere's Law}\end{aligned} \tag{4.30d}$$

where I trust you remember terms like "Gaussian surface" and "Amperian loop," that is, the "S" and "C", respectively in the integrals on the left side of (4.30).

Surface theorems can be used to turn each of the integrals on the left side into integrals over the Gaussian surface enclosing the volume, or Amperian loop enclosing the surface. For Gauss' Law, the charge enclosed by the Gaussian surface is just an integral over the volume of the charge density $\rho(\vec{r},t)$. The current enclosed by the Amperian loop is just the current density $\vec{j}(\vec{r},t)$ integrated over the surface it encloses. Therefore, these equations become

$$\vec{\nabla}\cdot\vec{E} = 4\pi\rho \qquad \text{Gauss' Law} \tag{4.31a}$$

$$\vec{\nabla}\cdot\vec{B} = 0 \qquad \text{Gauss' Law for Magnetism} \tag{4.31b}$$

$$\vec{\nabla}\times\vec{E} = -\frac{1}{c}\frac{\partial\vec{B}}{\partial t} \qquad \text{Faraday's Law} \tag{4.31c}$$

$$\vec{\nabla}\times\vec{B} = \frac{4\pi}{c}\vec{j} + \frac{1}{c}\frac{\partial\vec{E}}{\partial t} \qquad \text{Ampere's Law} \tag{4.31d}$$

This form of Maxwell's equations is much more amenable to studying the properties of charges and currents than is the integral form. It will also allow us to see immediately how they predict the existence of electromagnetic waves, which we will discuss in Section 4.3.4.2, and that electric charge is conserved, which we'll investigate now.

4.3.4.1 Conservation of electric charge

It is easy to show that (4.31) imply that electric charge is conserved. That is, it is easy to show that they lead to the continuity equation for the electric charge density

[3] I am writing these in Gaussian units, which are favored among physicists. Engineers typically use SI units, which is likely how you saw them in your first class.

$\rho(\vec{r},t)$ and current density $\vec{j}(\vec{r},t)$. First take the divergence of both sides of Ampere's Law. You know that the divergence of any curl is zero, so the left hand side must be zero. This gives us

$$0 = \frac{4\pi}{c}\vec{\nabla}\cdot\vec{j} + \frac{1}{c}\frac{\partial}{\partial t}\vec{\nabla}\cdot\vec{E}$$

Now use Gauss' Law to write $\vec{\nabla}\cdot\vec{E} = 4\pi\rho$. The overall factor of $4\pi/c$ cancels and you get

$$0 = \vec{\nabla}\cdot\vec{j} + \frac{\partial\rho}{\partial t}$$

which is just the continuity equation (4.29).

Maxwell's Equations imply that electric charge is conserved. This very important aspect of electrodynamics is often overlooked in classes on the subject.

4.3.4.2 Existence of electromagnetic waves

Maxwell's Equations predict the existence of electromagnetic waves. Mathematically, this means that Maxwell's Equations predict that there are forms of $\vec{E}(\vec{r},t)$ and $\vec{B}(\vec{r},t)$ that satisfy a partial differential equation known as the Wave Equation. We will look at how we go about solving this and similar equations in Sections 4.5 and 5.1, but for now, let's just go ahead and manipulate Maxwell's Equations to get the Wave Equation.

First, though, we will take a moment to prove a vector differential operator identity, namely

$$\vec{\nabla}\times(\vec{\nabla}\times\vec{A}) = \vec{\nabla}(\vec{\nabla}\cdot\vec{A}) - \vec{\nabla}^2\vec{A} \tag{4.32}$$

This is easy to do using the "δ, ε" technique of Section 4.1.2.1. Using the summation notation with our generic notation for Cartesian unit vectors, the gradient operator becomes

$$\vec{\nabla} = \hat{e}_i\frac{\partial}{\partial x_i}$$

where $x_1 = x$, $x_2 = y$, and $x_3 = z$. Since none of the $\hat{e}_i$ depend on position, we can just work with the components directly, and write

$$\begin{aligned}
[\vec{\nabla}\times(\vec{\nabla}\times\vec{A})]_i &= \varepsilon_{ijk}\frac{\partial}{\partial x_j}\left(\varepsilon_{kmn}\frac{\partial}{\partial x_m}A_n\right) = \varepsilon_{kij}\varepsilon_{kmn}\frac{\partial^2 A_n}{\partial x_j x_m} \\
&= (\delta_{im}\delta_{jn} - \delta_{in}\delta_{jm})\frac{\partial^2 A_n}{\partial x_j x_m} = \frac{\partial^2 A_j}{\partial x_j x_i} - \frac{\partial^2 A_i}{\partial x_j x_j} \\
&= \frac{\partial}{\partial x_i}\frac{\partial A_j}{\partial x_j} - \frac{\partial^2}{\partial x_j x_j}A_i = [\vec{\nabla}(\vec{\nabla}\cdot\vec{A}) - \vec{\nabla}^2\vec{A}]_i
\end{aligned}$$

where I have freely exchanged the order of differentiation when it suited me.

Equation (4.32) let's show that Maxwell's Equations predict the existence of electromagnetic waves. Let's see how this works. If we just talk about some region in space where there are no charges or currents, call it the "vacuum," then Maxwell's Equations become

$$\vec{\nabla}\cdot\vec{E}=0 \qquad \vec{\nabla}\cdot\vec{B}=0 \qquad \vec{\nabla}\times\vec{E}=-\frac{1}{c}\frac{\partial\vec{B}}{\partial t} \qquad \vec{\nabla}\times\vec{B}=\frac{1}{c}\frac{\partial\vec{E}}{\partial t}$$

Take the curl of the third equation, invoke (4.32), and use the first equation to get

$$\vec{\nabla}\times(\vec{\nabla}\times\vec{E})=\vec{\nabla}(\vec{\nabla}\cdot\vec{E})-\vec{\nabla}^2\vec{E}=-\vec{\nabla}^2\vec{E}=-\frac{1}{c}\frac{\partial}{\partial t}\vec{\nabla}\times\vec{B}$$

Finally, use the fourth equation to replace $\vec{\nabla}\times\vec{B}$ with the time derivative of $\vec{E}$. After a little rearranging, you find

$$\frac{1}{c^2}\frac{\partial^2\vec{E}}{\partial t^2}-\vec{\nabla}^2\vec{E}=0 \tag{4.33}$$

This partial differential equation is called the *Wave Equation*. We will discuss solutions to the wave equation in Section 5.1, but it describes a field $\vec{E}(\vec{r},t)$ with a shape that moves in time, unchanged, at speed c.

Equation (4.33) looks a little weird because it is a partial differential equation that is to be solved for a *vector* function $\vec{E}(\vec{r},t)$. Let's try writing $\vec{E}(\vec{r},t)=\hat{i}f(z,t)$, which describes an electric field that is "linearly polarized in the x-direction." Inserting this (4.33) gives us

$$\frac{1}{c^2}\frac{\partial^2 f}{\partial t^2}-\frac{\partial^2 f}{\partial z^2}=0$$

which is a bit more tractable. We will see in Section 5.1 that this equation describes a wave moving in the z-direction.

You will do much more with electromagnetic waves in your E&M courses. It all boils down, though, to using Maxwell's Equations to derive the relevant PDE's and then solving them, given initial and boundary conditions.

4.4 TWO IMPORTANT VECTOR FIELDS

Let's pause briefly to talk about two particular vector field forms that I'll call $\vec{B}(\vec{r})$ and $\vec{E}(\vec{r})$. Describing $\vec{B}(\vec{r})$ in Cartesian and *cylindrical* polar coordinates,

$$\vec{B}(\vec{r})=a\frac{-\hat{i}y+\hat{i}x}{x^2+y^2}=a\frac{\hat{\phi}}{\rho} \tag{4.34}$$

I will define the field $\vec{E}(\vec{r})$ in Cartesian and *spherical* polar coordinates as

$$\vec{E}(\vec{r})=a\frac{\hat{i}x+\hat{j}y+\hat{k}z}{(x^2+y^2+z^2)^{3/2}}=a\frac{\hat{r}}{r^2} \tag{4.35}$$

In both of these equations, a just represents some constant. Figure 4.7 shows vector plots of these two fields.

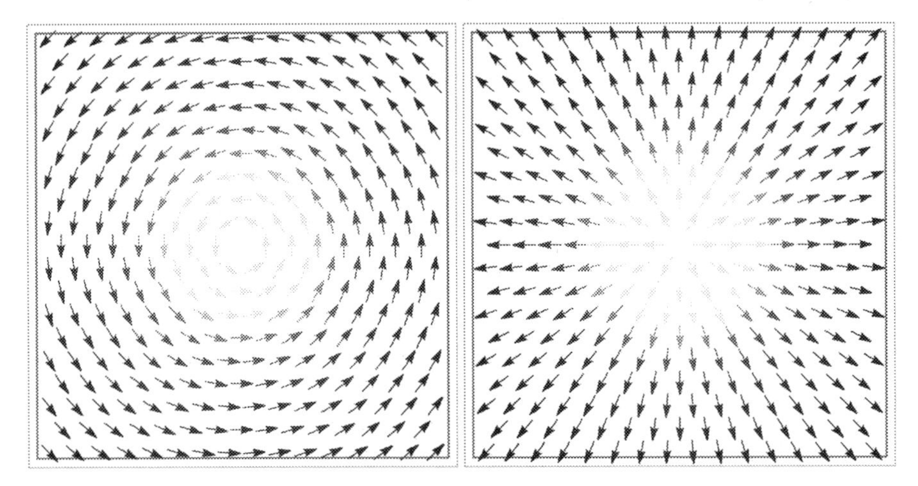

Figure 4.7 Vector plots of the fields given by (4.34) on the left, and by (4.35) on the right. The gray tone indicates the magnitude of the field, with the light tones representing larger values. It sure looks like (4.34) "curls" and (4.35) "diverges," but if you calculate the curl and divergence, respectively, you get zero in both cases.

I will leave it as a homework problem to calculate $\vec{\nabla} \times \vec{B}$ and $\vec{\nabla} \cdot \vec{E}$. Not to spoil the suspense, but you will find that the answer is zero in both cases. This is despite the fact that, from Figure 4.7, $\vec{B}(\vec{r})$ looks like it "curls" and $\vec{E}(\vec{r})$ very obviously "diverges."

In fact, if you try testing the surface theorems on these fields, you will get nonzero results! The easiest way to do this for $\vec{B}(\vec{r})$, is to use a circle C of radius R centered at the origin in the xy plane, and trace it counter clockwise. It is easy, then, to see that

$$\oint_C \vec{B}(\vec{r}) \cdot d\vec{\ell} = \oint_C a \frac{d\ell}{R} = 2\pi a$$

For $\vec{E}(\vec{r})$, use a sphere S of radius R centered on the origin to find

$$\oint \vec{E}(\vec{r}) \cdot d\vec{S} = \oint_S a \frac{dS}{R^2} = 4\pi a$$

This looks like violations of Stokes' Theorem and Gauss' Theorem, but in fact they are not. There is a subtlety that should be obvious when you think about it. It will take us a little while to get to the mathematics we need to write down what I'm talking about, but I wanted to get you thinking about it first.

4.5 PARTIAL DIFFERENTIAL EQUATIONS

Partial Differential Equations, or PDE's, are differential equations with more than one independent variable, so involve partial derivatives with respect to those variables. You will see them everywhere in physics, from electromagnetism, to quantum mechanics, and continuum mechanics. There is only enough time to barely have a discussion about PDE's in general, but we will highlight the most common techniques used to solve them.

Probably the most important difference, effectively, between PDE's and ODE's is that the boundary conditions are much more involved and can lead to much more general solutions. We will be dealing with linear PDE's only, so superposition will still be valid. *This will be key to determining linear combinations of solutions that satisfy boundary conditions.*

It's good to illustrate these points with a problem that looks simple, but shows that boundary conditions are critical to finding even the general form of a solution. Consider the PDE

$$\frac{\partial f}{\partial x} = x\frac{\partial f}{\partial y} \tag{4.36}$$

to be solved for the function $f(x,y)$. You can easily verify that the following are *all* solutions to this PDE:

$$\begin{aligned}
f_1(x,y) &= x^6 + 6x^4y + 12x^2y^2 + 8y^3 - 2 \\
f_2(x,y) &= \cos\left(x^2\right)\cos^2(y) - \cos\left(x^2\right)\sin^2(y) - 2\sin\left(x^2\right)\sin(y)\cos(y) \\
f_3(x,y) &= \frac{x^2 + 2y - 1}{2x^2 + 4y + 3}
\end{aligned}$$

That seems odd until you realize that all three are functions of $z = x^2 + 2y$. That is,

$$\begin{aligned}
f_1(x,y) &= z^3 - 2 \\
f_2(x,y) &= \cos z \\
f_3(x,y) &= \frac{z-1}{2z+3}
\end{aligned}$$

Indeed, for any function $g(z)$,

$$\frac{\partial g}{\partial x} = g'(z)\frac{\partial z}{\partial x} = 2xg'(z) \qquad \text{and} \qquad x\frac{\partial g}{\partial y} = xg'(z)\frac{\partial z}{\partial y} = 2xg'(z)$$

so any function $f(x,y) = g(z)$ solves the PDE. Which of these infinite possible forms of the solution you pick, though, will depend on what boundary conditions you need to satisfy.

In fact, first-order linear homogeneous PDE's of the form

$$p(x,y)\frac{\partial f}{\partial x} = q(x,y)\frac{\partial f}{\partial y} \tag{4.37}$$

can be solved by a prescription. Assuming a solution $f(x,y) = g(z)$ for some z that depends on x and y, that is $z = z(x,y)$, then

$$\begin{aligned}
df &= \frac{\partial f}{\partial x}dx + \frac{\partial f}{\partial x}dy = g'(z)\frac{\partial z}{\partial x}dx + g'(z)\frac{\partial z}{\partial y}dy \\
&= g'(z)\left[\frac{\partial z}{\partial x}dx + \frac{\partial z}{\partial y}dy\right] = g'(z)\,dz
\end{aligned}$$

This is really just a formal way of saying that if z doesn't change, even if x and y do change, then $f(x,y)$ doesn't change. Therefore, setting $df = 0$ and using (4.37), get

$$\frac{dy}{dx} = -\frac{\partial f/\partial x}{\partial f/\partial y} = -\frac{q(x,y)}{p(x,y)} \tag{4.38}$$

This expression can, in principle, be integrated to find a combination of x and y that is constant, and this determines z.

Let's try this out with (4.36), for which $p(x,y) = 1$ and $q(x,y) = x$. Then

$$\frac{dy}{dx} = -x \qquad \text{so} \qquad y = -\frac{1}{2}x^2 + c$$

where c is constant, and we recover what we discovered above with $z = 2c = x^2 + 2y$.

4.5.1 SEPARATION OF VARIABLES

Most if not all of the PDE's you'll need to solve in your studies of physics will be linear and second order, and most if not all of these can be solved using a technique called *Separation of Variables*. This section goes through the principles of this technique, and it will be applied in various places throughout the rest of this course. You will see it used often in your courses on electromagnetism and quantum mechanics.

Separation of variables turns a partial differential equation for a function in terms of two or three independent variables (like x, y, and z, or r, θ, and ϕ into a set of separate ordinary differential equations, one for each of the independent variables. Suppose we are looking to solve a PDE for a function $f(\vec{r}) = f(x,y,x)$. The first step is to write

$$f(x,y,z) = X(x)Y(y)Z(z)$$

and insert into the PDE. It is generally possible to arrange the terms in a way that some constant or constants can be used to isolate the different ODE's.

We will illustrate this idea in a moment, but first an important point: What coordinate system you use is probably most dependent on what are the boundary conditions. If you have $f(\mathbf{r})$ and its partial derivatives defined along the boundaries of a rectangle or rectangular box, then you likely want to use Cartesian coordinates. If they are defined along the boundaries of a sphere, then you are apt to use spherical coordinates.

Each problem is different, though, and you have to consider it carefully before you plow forward. Remember that you still have existence and uniqueness to help you out. Any way that you can find a solution to the PDE that satisfies the boundary conditions, will give you, essentially, the right solution.

So, instead of writing down any general rules, let's solve some specific problems so that you can see how this technique works. See also Section 5.1.

4.5.1.1 Example: Solution on a square

Let's solve the partial differential equation

$$\frac{\partial^2 f}{\partial x^2} = \frac{\partial^2 f}{\partial y^2}$$

for the function $f(x,y)$ subject to the boundary conditions

$$f(0,y) = f(a,y) = f(x,0) = f(x,a) = 0$$

where a is a positive constant. That is, the region of validity of the solution is the square of side length a in the first quadrant.

If you think about it, there is a more or less obvious solution to the differential equation. If $f(x,y) = g(z)$ with $z = x \pm y$ for any function $g(z)$, then the PDE is satisfied. In fact, a general solution would be

$$f(x,y) = g_1(x-y) + g_2(x+y)$$

where g_1 and g_2 are arbitrary functions. However, it's not so obvious how to choose these functions so that the boundary conditions are satisfied.

For this, we turn to a solution that makes use of Separation of Variables. As suggested above, we write $f(x,y) = X(x)Y(y)$ which gives

$$\frac{d^2X}{dx^2}Y = X\frac{d^2Y}{dy^2} \qquad \text{so} \qquad \frac{1}{X}\frac{d^2X}{dx^2} = \frac{1}{Y}\frac{d^2Y}{dy^2}$$

which is a peculiar type of equation. The left side depends only on x, while the right side depends only on y. Nevertheless, the two sides have to be equal to each other! The only way this happens if both sides are equal to a constant.

Let's call this constant $-k^2 < 0$. (You'll see shortly that this choice for the constant allows us to satisfy the boundary conditions.) The PDE has now split into two ODE's, as advertised. The two ODE's are

$$X''(x) = -k^2X(x) \qquad \text{and} \qquad Y''(y) = -k^2Y(y)$$

whose solutions are now very well-known to you, namely

$$X(x) = A_x\cos kx + B_x\sin kx \qquad \text{and} \qquad Y(y) = A_y\cos ky + B_y\sin ky$$

where A_x, B_x, A_y, and B_y are constants. The boundary conditions imply that

$$X(0) = 0 = Y(0) \qquad \text{and} \qquad X(a) = 0 = Y(a)$$

The first pair of equations say that $A_x = 0 = A_y$, so the solution to the PDE has the form

$$f(x,y) = C\sin kx\sin ky$$

where $C = B_xB_y$. In order for $f(a,y) = 0 = f(x,a)$, and to avoid the trivial solution with $C = 0$, we need to set $k = n\pi/a$ were $n \in \mathbb{Z}^+$ is a positive integer. (A negative integer would give a redundant function, and $n = 0$ leads to the trivial solution.) Therefore

$$f(x,y) = C\sin\left(\frac{n\pi x}{a}\right)\sin\left(\frac{n\pi y}{a}\right)$$

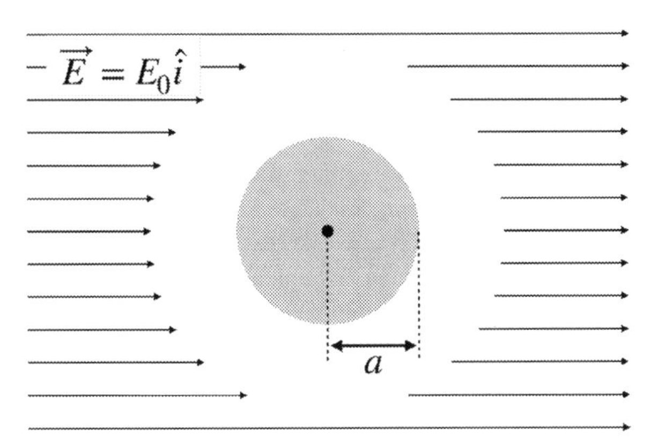

Figure 4.8 A uniform electric field with magnitude E_0 in the x-direction permeates all space, and then a conducting rod of radius a is moved into the space, perpendicular to the field. The problem is to find how the field is modified in the region of the rod.

where the constant C would have to be determined by some additional constraint. If you recall the trigonometric identity

$$\sin\alpha\sin\beta = \frac{1}{2}\left[\cos(\alpha-\beta)-\cos(\alpha+\beta)\right]$$

then you see that, indeed, $f(x,y)$ is a linear combination of functions of $x \pm y$.

This worked out nicely because I picked boundary conditions that were amenable to a solution with separation of variables. Luckily, most of the physics problems you will face work out nicely like this. For other cases, not uncommon in various types of modeling of systems, you may need to resort to numerical solutions.

4.5.1.2 Example: Electrostatics in two dimensions

Imagine an electric field $\vec{E} = E\hat{i}$ in the x-direction that permeates all space. Now imagine that a long circular conducting rod of radius a is inserted with its axis in the z-direction. How does the electric field change to accommodate the presence of the rod? The situation is pictured in Figure 4.8.

The first thing to realize is that this is a problem in electro*statics*, so there is no time dependence. Therefore, Faraday's Law (4.31c) says that $\vec{\nabla}\times\vec{E} = 0$, and we can write $\vec{E}(\vec{r})$ as the gradient of some scalar function. For reasons you'll learn in a course in electromagnetism, we do this with a minus sign, that is

$$\vec{E}(\vec{r}) = -\vec{\nabla}V(\vec{r})$$

where $V(\vec{r})$ is called the electrostatic potential. Since we are concerned with the field outside the rod, where there are no charges, we use Gauss' Law (4.31a) to write $\vec{\nabla}\cdot\vec{E} = -\vec{\nabla}\cdot\vec{\nabla}V = -\vec{\nabla}^2V = 0$. In other words, to find $\vec{E}$, we want to solve

$$\vec{\nabla}^2V(\vec{r}) = 0$$

which is known as Laplace's Equation.

The symmetry of this problem clearly suggests that we should solve this partial differential equation using cylindrical coordinates. That is, we use (4.26b) to write

$$\frac{1}{\rho}\frac{\partial}{\partial\rho}\left(\rho\frac{\partial V}{\partial\rho}\right)+\frac{1}{\rho^2}\frac{\partial^2 V}{\partial\phi^2}+\frac{\partial^2 V}{\partial z^2}=0$$

Assuming that our rod is infinitely long, there will be no z-dependence, so we have $V(\vec{r})=V(\rho,\phi)$ which we write as $V(\rho,\phi)=R(\rho)\Phi(\phi)$ to approach the problem using separation of variables, and we arrive at

$$\frac{1}{R}\rho\frac{d}{d\rho}\left(\rho\frac{dR}{d\rho}\right)=-\frac{1}{\Phi}\frac{d^2\Phi}{d\phi^2} \tag{4.39}$$

where, again, we argue that each side of the equation must equal the same constant, in order for the equation to be satisfied for all ρ and ϕ.

We proceed, as always, by incorporating physical insight while guided by existence and uniqueness of the solution. We will want to enforce $\Phi(\phi+2\pi)=\Phi(\phi)$, so it makes sense to set each side of the equation equal to n^2, where n is a positive constant. Thus the solutions for $\Phi(\phi)$ will be proportional to $\cos(n\phi)$ or $\sin(n\phi)$. This choice leads to the differential equation for $R(\rho)$

$$\rho\frac{d}{d\rho}\left(\rho\frac{dR}{d\rho}\right)=n^2R \qquad \text{or} \qquad \rho^2\frac{d^2R}{d\rho^2}+\rho\frac{dR}{d\rho}-n^2R=0$$

which is just a form of the Euler Equation (3.22). Writing $R(\rho)=\rho^s$, we find

$$s(s-1)+s-n^2=s^2-n^2=0$$

so $s=\pm n$. Putting this all together, we see that $V(\rho,\phi)=R(\rho)\Phi(\phi)$ is a linear combination of ρ^n or ρ^{-n} times $\cos(n\phi)$ or $\sin(n\phi)$, with potentially different coefficients for all positive integers n, summed over n.

Of course, we use boundary conditions to sort through all these possibilities. We know that for $\rho\to\infty$, we must have $E(\vec{r})\to E_0\hat{i}$ so $V(\rho,\phi)\to -E_0x=-E_0\rho\cos\phi$. This tells us that $n=1$, and also fixes the coefficient of the term proportional to ρ. Therefore, we know that

$$V(\rho,\phi)=-\left(E_0\rho+\frac{A}{\rho}\right)\cos\phi$$

In order to find the constant A, we impose the boundary condition on the surface of the rod, which we will assume is made from a perfect conductor. In this case, the tangential component of the electric field at the surface must be zero. (This is because the field inside the conductor must be zero and the tangential component of the field must be continuous across the surface, which follows from $\vec{\nabla}\times\vec{E}=0$.) Setting the tangential component of the field at $\rho=a$, given by the negative of the ϕ-component of the gradient of $V(\rho,\phi)$, to zero, we have

$$-\left.\frac{1}{\rho}\frac{\partial V}{\partial\phi}\right|_{\rho=a}=-\frac{1}{a}\left(E_0a+\frac{A}{a}\right)\sin\phi=0$$

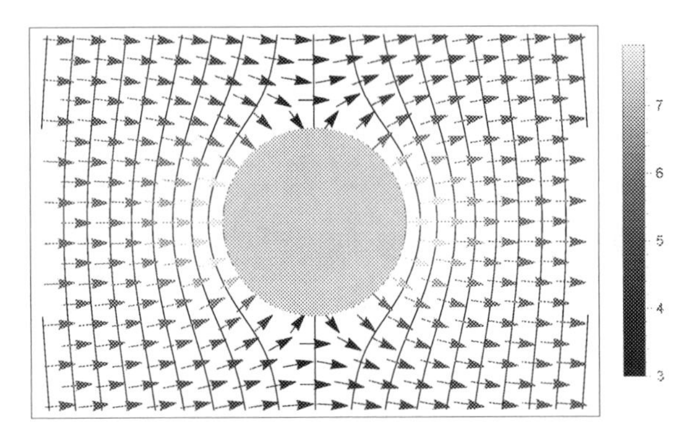

Figure 4.9 The solution to the problem depicted in Figure 4.8. The electric potential is plotted as a contour plot, along with a vector plot of the resulting electric field.

which tells us that $A = -E_0 a^2$. Therefore, the electric potential is

$$V(\rho,\phi) = -E_0\left(\rho - \frac{a^2}{\rho}\right)\cos\phi$$

It is worth your time to take a minute and confirm that this form satisfies Laplace's equation, as well as the boundary conditions.

Figure 4.9 shows the solution as a contour plot of $V(\rho,\phi)$ and and a vector plot of $\vec{E} = -\vec{\nabla}V$. Compare this to Figure 4.8. Note that the electric field vectors are perpendicular to the surface of the rod at $\rho = a$.

We know that this is the right solution, thanks to uniqueness, but you might have noticed that I slipped something past you. We chose $n = 1$ based on the boundary condition for $\rho \to \infty$, but in principle there could have been an infinite sum over n for the $1/\rho$ form of $R(\rho)$. In fact, as we'll see in Section 5.2, the coefficients for $n \geq 2$ must all be zero.

4.5.1.3 Example: Heat conduction in one dimension

A common problem in the physical and life sciences, including engineering, is the diffusion of heat (or equivalent entity) through a medium over time. This process is governed by a partial differential equation of the form

$$\vec{\nabla}^2 f = k\frac{\partial f}{\partial t} \tag{4.40}$$

where $f(\vec{r},t)$ is the temperature field and k is a positive constant, called the *diffusivity*. This equation implies that there is no heat source in the medium, but that its internal energy can only get redistributed by the transfer of heat. We refer to (4.40) as the *heat equation* or *diffusion equation*.

In one spatial dimension x, (4.40) becomes

$$\frac{\partial^2 f}{\partial x^2} = k\frac{\partial f}{\partial t} \tag{4.41}$$

which would govern, for example, the temperature distribution $f(x,t)$ for position along a rod as a function of time. If the rod has length ℓ, its ends are held at fixed temperatures, and its initial temperature distribution is some function $u(x)$ for $0 \leq x \leq \ell$, then you have a prototypical PDE boundary and initial value problem to solve.

We should think for a minute about the physical system (4.41) would describe. After some time, we expect that the rod would come to thermal equilibrium with its "surroundings," namely the source or sink of heat to which each endpoint is attached. In equilibrium, the temperature $f(x,t)$ would no longer change with time, and the right side of (4.41) would be zero. The resulting (ordinary) differential equation is trivial to solve. You get

$$f(x,t \to \infty) = ax + b$$

where the constants a and b are set by the temperatures at the ends of the rod. Something in the time dependence of $f(x,t)$ must have the time dependence disappear after a long time.

A natural approach to solving (4.41) for the temperature of a rod, is to use Separation of Variables and write $f(x,t) = X(x)T(t)$. In this case (4.41) becomes

$$\frac{1}{X}\frac{d^2X}{dx^2} = k\frac{1}{T}\frac{dT}{dt}$$

and we need to set both sides equal to a constant. If we make this constant negative, for example $-\alpha^2$, then $T(t)$ will be a decaying exponential, which sounds like it fits the bill for a disappearing time dependence.

We would also find that $X(x)$ is a linear combination of sines and cosines. Enforcing fixed temperatures at the end of the rod will have us write the argument of the sines and cosines as some integer times π/ℓ. In order to satisfy the initial condition that $f(x,0 = u(x)$ for an arbitrary $u(x)$, we would have to find the linear combinations of all of the n-valued sines and cosines that give you $u(x)$.

This process of adding up sines and cosines to give you some arbitrary function was invented in the early 19th century by Joseph Fourier, and is the subject of Chapter 5. According to lore, Fourier developed this technique to solve the problem of how to best dissipate heat in the cannons of Napoleon Bonaparte's army. We'll use a more traditional approach to develop Fourier Series, and then the Fourier Transform, namely standing waves on a string.

EXERCISES

4.1 *Derive the unit vectors $\hat{r}$, $\hat{\theta}$, $\hat{\phi}$ in spherical polar coordinates in terms of the Cartesian unit vectors $\hat{\imath}$, $\hat{\jmath}$, $\hat{k}$ and the spherical polar coordinates r, θ, and ϕ. Your starting point should be the transformation equations that give you the Cartesian coordinates x, y, and z in terms of the spherical coordinates.*

4.2 *The kinetic energy of a particle of mass m is $K = mv^2/2 = m\vec{v}\cdot\vec{v}/2$, where $\vec{v} = d\vec{r}/dt$ is the particle's velocity vector. Derive an expression for K in terms of the spherical coordinates r, θ, and ϕ and their rates of change $\dot{r}$, $\dot{\theta}$, and $\dot{\phi}$ with respect* *to time. Simplify your result as much as possible. You can carry all this out with the chain rule and the relevant transformation equations, but there is also a much simpler way.*

4.3 *Find an expression for the square of the magnitude of the cross product* $|\vec{A}\times\vec{B}|^2$ *in terms of the magnitudes of* $\vec{A}$ *and* $\vec{B}$ *and their dot product* $\vec{A}\cdot\vec{B}$*, in two different ways:*

(a) *Directly from the definitions of the magnitudes of* $|\vec{A}\times\vec{B}|$ *and* $\vec{A}\cdot\vec{B}$.
(b) *Using components and the summation convention, along with the relationship between the totally antisymmetric symbol and the Kronecker delta.*

4.4 *Consider two lines in a plane, described by* $y = m_1x + b_1$ *and* $y = m_2x + b_2$. *Show that if the two lines are perpendicular at their intersection, then* $m_1m_2 = -1$.

4.5 *Find the equation of a plane which contains the point* $(x,y,z) = (1,-2,5)$ *and which is perpendicular to a vector pointing from the origin into the first quadrant and which makes equal angles with the x, y, and z axes. Write your equation in the form* $Ax + By + Cz = D$*, where A, B, C, and D have numerical values.*

4.6 *A line in space passes through the origin and is at an angle of* 45° *with respect to the positive z-axis and is at equal angles with respect to the positive x- and y-axes. Find the coordinates of the intersection point of this line with the plane from the previous problem.*

4.7 *For spatial vectors* $\vec{A}$, $\vec{B}$, $\vec{C}$, *and* $\vec{D}$, *prove that*

$$(\vec{A}\times\vec{B})\cdot(\vec{C}\times\vec{D}) = (\vec{A}\cdot\vec{C})(\vec{B}\cdot\vec{D}) - (\vec{A}\cdot\vec{D})(\vec{B}\cdot\vec{C})$$

I think the easiest way to do this is to write the vectors in terms of their components and make use of the Kronecker δ *and the Levi-Civita symbol, and their properties. Now use this result to find an expression for* $|\vec{A}\times\vec{B}|^2$ *in terms of the magnitudes of* $\vec{A}$ *and* $\vec{B}$ *and their dot products. Explicitly show that this is the same as the geometric definition of the magnitude of the cross product.*

4.8 *For a particle of mass m moving in a plane located at position* $\vec{r}(t)$*, find an expression for the kinetic energy*

$$K = \frac{1}{2}m\left(\frac{d\vec{r}}{dt}\right)^2$$

in terms of plane polar coordinates r and ϕ. *Do this explicitly by writing* $\vec{r}$ *first in terms of Cartesian coordinates x and y, convert to polar coordinates, and then take derivatives.*

4.9 *These are simple practice exercises for vector field operators.*

(a) *Calculate the gradient of the scalar field* $f(x,y,z) = xyz$.

(b) *Calculate the divergence of the vector field* $\vec{v}(x,y,z) = \hat{\imath}x + \hat{\jmath}z - \hat{k}y$.
(c) *Calculate the curl of the vector field* $\vec{v}(x,y,z) = \hat{\imath}x + \hat{\jmath}z - \hat{k}y$.

4.10 *Show that the gradient operator in spherical coordinates is given by*

$$\vec{\nabla} = \hat{r}\frac{\partial}{\partial r} + \hat{\theta}\frac{1}{r}\frac{\partial}{\partial \theta} + \hat{\phi}\frac{1}{r\sin\theta}\frac{\partial}{\partial \phi}$$

You can do this using the transformation equations and the chain rule for partial derivatives, but you don't need to do nearly that much work. Think of the gradient as a "directional derivative" as discussed in the text, and the unit vectors $\hat{r}$, $\hat{\theta}$, $\hat{\phi}$ *i in terms of* $\hat{\imath}$, $\hat{\jmath}$, $\hat{k}$ *to write down what is the change* $d\vec{r}$ *for each of the three orthogonal directions in spherical coordinates.*

4.11 *Derive the Laplacian in plane polar coordinates, i.e. cylindrical coordinates with no z-component. That is, show that*

$$\begin{aligned}\vec{\nabla}^2 f(r,\phi) = \vec{\nabla}\cdot\vec{\nabla} f(r,\phi) &= \left(\hat{r}\frac{\partial}{\partial r} + \hat{\phi}\frac{1}{r}\frac{\partial}{\partial \phi}\right)\cdot\left(\hat{r}\frac{\partial f}{\partial r} + \hat{\phi}\frac{1}{r}\frac{\partial f}{\partial \phi}\right) \\ &= \frac{1}{r}\frac{\partial}{\partial r}\left(r\frac{\partial f}{\partial r}\right) + \frac{1}{r^2}\frac{\partial f}{\partial \phi^2}\end{aligned}$$

Don't forget that you need to take into account that $\hat{r}$ *and* $\hat{\phi}$ *depend explicitly on* ϕ. *Writing these unit vectors in terms of* $\hat{\imath}$ *and* $\hat{\jmath}$ *is probably the easiest way to do this.*

4.12 *For a scalar field* $f(\vec{r})$ *over some volume V in three-dimensional space, prove that*

$$\int_V \vec{\nabla} f\, dV = \oint_S f\, d\vec{S}$$

where S is the surface enclosing V. You can use the "cut the volume up into little bricks" approach we used in class, or you can try inventing a vector field $\vec{A} = f\vec{C}$ *where* $\vec{C}$ *is some arbitrary constant vector, and then use a different surface theorem.*

4.13 *A force field* $\vec{F}(\vec{r})$ *is said to be "conservative" if the work* $W = \int_1^2 \vec{F}(\vec{r})\cdot d\vec{r}$ *done between any two points in space 1 and 2 is independent of the path taken between these two points. Show that any force field of the form* $\vec{F}(\vec{r}) = f(r)\hat{r}$, *where r is the distance from the origin and* $\hat{r}$ *is the radial unit vector, is conservative.*

4.14 *Calculate the curl of the following vector field in both Cartesian coordinates and cylindrical polar coordinates:*

$$\vec{A}(\vec{r}) = -\frac{\hat{\imath}y - \hat{\jmath}x}{x^2 + y^2} = \frac{\hat{\phi}}{r}$$

Now calculate directly the line integral $\oint \vec{A}\cdot d\vec{\ell}$ *around a closed circle of radius R in the xy plane, centered at the origin. (I suggest you do this with the polar coordinate expression.) Can you reconcile these two seemingly inconsistent results?*

4.15 *Calculate the divergence of the following vector field in both Cartesian coordinates and spherical polar coordinates:*

$$\vec{A}(\vec{r}) = \frac{\hat{i}x + \hat{j}y + \hat{k}z}{(x^2+y^2+z^2)^{3/2}} = \frac{\hat{r}}{r^2}$$

Now calculate directly the surface integral $\oint \vec{A} \cdot d\vec{S}$ *around a sphere of radius* R*, centered at the origin. (I suggest you do this with the polar coordinate expression.) Can you reconcile these two seemingly inconsistent results?*

4.16 *Find the solution* $u(x,y)$ *for the partial differential equation and boundary condition*

$$\frac{\partial u}{\partial x} = -\frac{\partial u}{\partial y} \qquad u(x,0) = 2x$$

4.17 *Prove the "chain rule" for the divergence operator, namely for a scalar field* $f(\vec{r})$ *and a vector field* $\vec{A}(\vec{r})$*,*

$$\vec{\nabla} \cdot (f\vec{A}) = \vec{\nabla} f \cdot \vec{A} + f\vec{\nabla} \cdot \vec{A}$$

4.18 *The time-dependent Schrödinger Equation in three dimensions is*

$$-\frac{\hbar^2}{2m}\vec{\nabla}^2\psi(\vec{r},t) + V(\vec{r})\psi(\vec{r},t) = i\hbar\frac{\partial \psi}{\partial t}$$

where $V(\vec{r})$ *is a potential energy function and* $\psi(\vec{r},t)$ *is called the "wave function." Show that this equation implies that* $\rho(\vec{r},t) = \psi^*\psi$ *is a conserved density if its current density is given by* $j(\vec{r},t) = \hbar\,\mathrm{Im}(\psi^*\vec{\nabla}\psi)/m$*.*

4.19 *A magnetic field* $\vec{B}(r,\phi) = \hat{\phi}\, B_0\, (r/a)^2 \cos^2\phi$ *where* r *and* ϕ *are the polar coordinates in the* (x,y) *plane, and* B_0 *is a constant. Find the total enclosed current passing through a circle of radius* a *in the* (x,y) *plane centered at the origin. Do the necessary line integral directly, and compare to the result you get using Stokes' Theorem.*

4.20 *An electric field* $\vec{E}(r,\theta,\phi) = \hat{r}\, E_0\, (r/a) \cos^2\theta$ *where* r*,* θ*, and* ϕ *are the usual spherical coordinates, and* E_0 *is a constant. Find the total enclosed charge contained in a sphere of radius* a *centered at the origin. Do the necessary surface integral directly, and compare to the result you get using Gauss' Theorem.*

4.21 *Find the solutions* $u(x,y)$ *to the partial differential equation*

$$x\frac{\partial u}{\partial x} - 2y\frac{\partial u}{\partial y} = 0$$

separately for each of the following two boundary conditions:

(a) $u(x,y) = 2y+1$ *along the line* $x = 1$

(b) $u(1,1) = 4$, *that is, a single point. You may want to start by looking for a solution* $u(x,y) = f(p)$ *where* $p = p(x,y)$. *One way to do this (other than just guessing outright) is to relate the given differential equation to* $dp = 0$. *That is, the differential equation should be satisfied if* $dp = 0$ *as x and y change.*

4.22 *Use the "Separation of Variables" approach to solve the partial differential equation*

$$4\frac{\partial^2 u}{\partial x^2} = \frac{\partial u}{\partial t}$$

for the function $u(x,t)$ *with the initial condition* $u(x,0) = \sin(\pi x/2)$ *and boundary conditions* $u(2,t) = u(0,t) = 0$.

4.23 *Look for a solution to the Helmholtz Equation* $\vec{\nabla}^2 f(r,\phi) + k^2 f(r,\phi) = 0$ *in plane polar coordinates by writing* $f(r,\phi) = R(r)\Phi(\phi)$. *Now insist that* $\Phi(\phi + 2\pi) = \Phi(\phi)$, *that is* $\Phi(\phi)$ *must be "single valued," and show that solutions for* $R(r)$ *must be Bessel Functions* $J_m(kr)$ *of integer order m.*

4.24 *Look for a solution to the Helmholtz Equation* $\vec{\nabla}^2 f(r,\theta) + k^2 f(r,\theta) = 0$ *in spherical polar coordinates, where the solution has no explicit dependence on the azimuthal angle* ϕ, *by writing* $f(r,\phi) = R(r)\Theta(\theta)$. *Show that the solution has the form* $R(r) = j_\ell(kr)$ *and* $\Theta(\theta) = P_\ell(\cos\theta)$ *for some non-negative integer* ℓ, *where* $j_\ell(\rho)$ *is a solution to the spherical Bessel equation and* $P_\ell(x)$ *is the Legendre polynomial of degree* ℓ.

5 Fourier Analysis

Fourier Analysis is based on the fact – which we won't prove formally – that pretty much any function can be represented by an infinite sum, or perhaps an integral, of sines and cosines. Although applicable to any number of physical problems (for example, heat conduction, as mentioned in Section 4.5.1.3), we will illustrate it using solutions to the Wave Equation.

5.1 WAVES ON A STRETCHED STRING

We are going to start this discussion by considering the motion of a string, stretched tightly. We'll say the string is in the horizontal direction and moves only in the vertical direction, because it never bends by very much, but gravity is irrelevant. We are only going to care about the vertical motion due to the tension in the string. What we'll find is that the motion of the string is governed by the same partial differential equation that governs electromagnetic waves, for example (4.33).

5.1.1 DERIVATION OF THE EQUATION OF MOTION

Imagine a string stretched across space. We will derive an equation that governs the up-and-down motion of the string, assuming that it never bends by very much. We will do this by considering tiny piece of the string and applying Newton's Second Law. The only forces on this piece of string that will concern us are the tensions on the piece from each of its two ends.

Figure 5.1 shows a small piece of stretched string. We measure the position along the horizontal direction as x, and the shape of the string at any time t is $u(x,t)$. We assume that θ is always very small so that the string only moves vertically.

In order to apply Newton's Second Law, let's first analyze the vertical forces on the string. These are

$$-T\sin\theta \approx -T\theta \approx -T\tan\theta = -T\left.\frac{\partial u}{\partial x}\right|_{x}$$

on the left, and

$$T\sin\theta \approx T\theta \approx T\tan\theta = T\left.\frac{\partial u}{\partial x}\right|_{x+\Delta x}$$

on the right. Therefore, the sum of forces acting on this small piece of string is

$$\sum F_y = -T\left.\frac{\partial u}{\partial x}\right|_{x} + T\left.\frac{\partial u}{\partial x}\right|_{x+\Delta x} = T\left[\left.\frac{\partial u}{\partial x}\right|_{x+\Delta x} - \left.\frac{\partial u}{\partial x}\right|_{x}\right]$$

Now if we let μ be the linear mass density of the string, then the mass of this small piece is $\mu\Delta x$, and Newton's Second Law for the vertical motion of the piece of

DOI: 10.1201/9781003355656-5

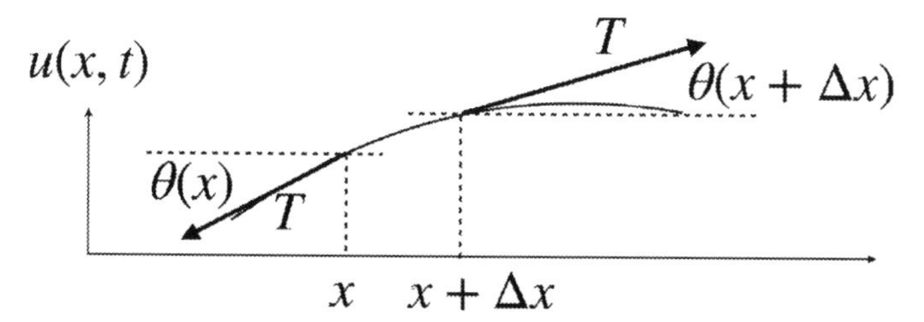

Figure 5.1 A small piece of a string, located between x and $x+\Delta x$, acted on by the tension forces from each end of the small piece. We use θ to measure the angle of the string with respect to the horizontal at each end.

string is

$$\mu \Delta x \frac{\partial^2 u}{\partial t^2} = T \left[\left.\frac{\partial u}{\partial x}\right|_{x+\Delta x} - \left.\frac{\partial u}{\partial x}\right|_{x} \right] \qquad \text{or} \qquad \frac{\mu}{T}\frac{\partial^2 u}{\partial t^2} = \frac{1}{\Delta x}\left[\left.\frac{\partial u}{\partial x}\right|_{x+\Delta x} - \left.\frac{\partial u}{\partial x}\right|_{x} \right]$$

It is easy to see that the constant μ/T on the left has dimensions of $1/(\text{velocity})^2$. That is

$$\frac{[\mu]}{[T]} = \frac{ML^{-1}}{MLT^{-2}} = \frac{1}{(L/T)^2}$$

So, let's write $v^2 = T/\mu$. Also, as the piece of string gets smaller and smaller, that is $\Delta x \to 0$, the right side just becomes the second partial derivative with respect to x. In other words, the motion of the string is governed by the partial differential equation

$$\frac{1}{v^2}\frac{\partial^2 u}{\partial t^2} = \frac{\partial^2 u}{\partial x^2} \qquad \text{or} \qquad \frac{\partial^2 u}{\partial x^2} - \frac{1}{v^2}\frac{\partial^2 u}{\partial t^2} = 0 \tag{5.1}$$

This is known as the *Wave Equation* and it shows up in many different areas of physics. We have already seen that it is implied by Maxwell's Equations in Section 4.3.4.2.

5.1.2 GENERAL SOLUTION OF THE WAVE EQUATION

There are many different forms for the function $u(x,t)$ that solves (5.1) that depend on the initial and boundary conditions. However, we can immediately see a specific general form of the solution that gives us good intuition as to what's going on.

First, realize that (5.1) is a linear PDE. That means that if $u_1(x,t)$ and $u_2(x,t)$ are both solutions, then any linear combination $c_1u_1(x,t)+c_2u_2(x,t)$ is also a solution.

Now it is not hard to see that any function of $z = x - vt$ is a solution to (5.1). That is, $u(x,t) = f(z) = f(x-vt)$ will be a solution for any (differentiable) function $f(z)$. The same is true for any function $g(x+vt)$. In other words, the general solution to (5.1) is any function of the form

$$u(x,t) = f(x-vt) + g(x+vt) \tag{5.2}$$

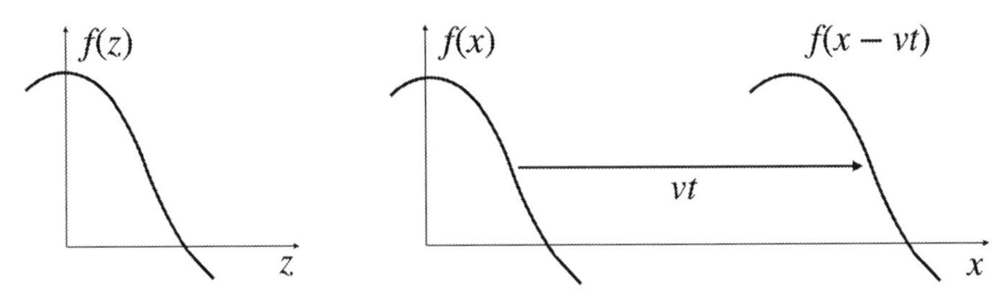

Figure 5.2 Motion of a string according to (5.1) for the part of the general solution (5.2) that is $f(x-vt)$. The left shows an arbitrary shape $f(z)$. The right shows $f(z) = f(x-vt)$ for $t = 0$, that is $z = x$, and also for arbitrary $t > 0$. The effect of a finite t is to translate the shape to the right by an amount vt. (This drawing ignores the fact that we've assumed that the angle of the string with the horizontal is always small.)

This general form has a lovely physical interpretation. Consider first the motion in time of the function $f(x-vt)$, shown in Figure 5.2. Whatever the shape is at $t = 0$, that is $f(x)$, it is reproduced exactly at a finite time t except that it is translated to the right by an amount vt. The speed at which it moves is the distance divided by the time, that is $vt/t = v$.

In other words, the first term in the solution (5.2) represents a "wave moving to the right with speed v." The second term, that is $g(x+vt)$ represents, similarly, a "wave moving to the left with speed v."

It is not hard to show that if you start at time $t = 0$ with some arbitrary string shape $f(x)$, with the string at rest, then the solution is that the shape splits into two pieces, one moving to the right and the other moving to the left. This is left as an exercise.

Another easy solution to the wave equation is for a string that is fixed at one end, or allowed to move freely at one end. These cases correspond to the boundary conditions that $u(0,t) = 0$ or $\partial u(x,t)/\partial x|_{x=0} = 0$, assuming the end to be at $x = 0$. In each case, you find that the wave is "reflected" from the end, but the characteristics of the reflection are different for the two cases. This is also left as an exercise.

5.2 STANDING WAVES

Now we are going to investigate solutions to (5.1) with some specific boundary conditions, namely that the ends of the string are fixed and cannot vibrate. The result is a phenomenon that we refer to as *standing waves*.

Let's set up the problem with the ends of the string at $x = 0$ and $x = L$. Our goal is to find a solution to (5.1) subject to the boundary conditions

$$u(0,t) = 0 \qquad \text{and} \qquad u(L,t) = 0 \tag{5.3}$$

We will worry about the initial conditions later. A good approach to solving this problem is the technique of separation of variables, discussed in Section 4.5.1. This implies that we want to write the solution as

$$u(x,t) = X(x)T(t) \qquad \text{where} \qquad X(0) = 0 \qquad \text{and} \qquad X(L) = 0$$

and insert this into (5.1). This leads us to

$$\frac{1}{v^2}\frac{1}{T}\frac{d^2T}{dt^2} = \frac{1}{X}\frac{d^2X}{dx^2} = -k^2 \tag{5.4}$$

We choose the constant $-k^2 < 0$ because, as we'll see in a moment, this allows us to easily meet the boundary conditions.

The solution to (5.4) for $X(x)$ is best written as

$$X(x) = A\cos(kx) + B\sin(kx) \tag{5.5}$$

Enforcing $X(0) = 0$ gives $A = 0$. Enforcing $X(L) = 0$ gives $kL = n\pi$, where n is some integer. We can also write down the solution for $T(t)$ in the same way, that is

$$T(t) = C\cos(kvt) + D\sin(kvt) \tag{5.6}$$

where C and D are determined from the initial conditions.

At this point, we will make our lives easier by assuming that the string is initially at rest. Physically, this means that we "pluck" the string by giving it some initial shape $f(x)$ and then releasing it. In other words, we start with

$$u(x,0) = f(x) \qquad \text{and} \qquad \left.\frac{\partial u}{\partial t}\right|_{t=0} = 0 \tag{5.7}$$

The second equation implies that $\dot{T}(0) = 0$, that is, from (5.6),

$$-kvC\sin(0) + kvD\cos(0) = kvD = 0$$

which means that $D = 0$ in (5.6). Therefore, the solution to (5.1) takes the form

$$u_n(x,t) = B_n \sin\left(\frac{n\pi x}{L}\right)\cos\left(\frac{n\pi vt}{L}\right) \tag{5.8}$$

The notation here is important. We have the freedom to choose any positive integer n[1] This means that we have an infinite number of possible solutions, which we label as $u_n(x,t)$, each with a (so far) arbitrary constant B_n out front.

The solutions (5.8) should familiar to you from your first physics course. These are the so-called "fundamental modes" for standing waves on a string, and might be more familiar to you if we write them as

$$u_n(x,t) = B_n \sin\left(2\pi\frac{x}{\lambda_n}\right)\cos\left(2\pi f_n t\right) \tag{5.9}$$

where $\lambda_n \equiv 2L/n$ is the "wavelength" of the mode, and $f_n \equiv nv/2L$ is the "frequency" of the mode. The first three fundamental notes are plotted in Figure 5.3. Note that

[1] Choosing $n = 0$ leads to $u(x,t) = 0$ so we ignore that. Choosing $n < 0$ is the same choice as for $n > 0$ with just a change in sign of the constant out front, so we ignore that, too.

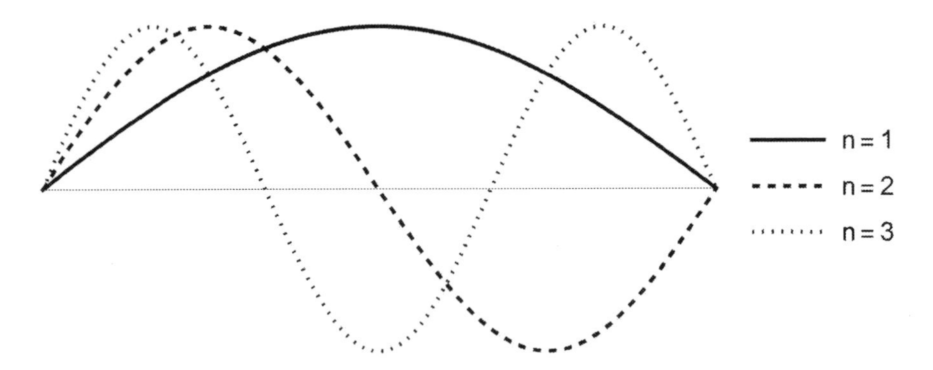

Figure 5.3 Shape of the first three fundamental modes of a stretched string. It is clear that the wavelengths are $2L$, L, and $2L/3$, where L is the length of the (unstretched) string.

$\lambda_n f_n = v$, that is, the product of the wavelength and frequency is the speed of the wave.

We now need to confront the initial condition $u(x,0) = f(x)$. At first glance, something seems very wrong. How can we have the initial condition be some arbitrary function $f(x)$ when (5.8) tells us that it has to be some function $\sin(n\pi x/L)$? We realize, though, that (5.1) is a linear differential equation so the principle of superposition holds. (Recall the discussion in Section 3.1.1.) This means any sum of solutions is still a solution, so we write

$$f(x) = \sum_{n=1}^{\infty} B_n \sin\left(\frac{n\pi x}{L}\right) \tag{5.10}$$

where the job is now to find the B_n. (Sometimes this is called the *Fourier Sine Series*.)

There is an easy way to find the B_n by exploiting the orthogonality of the sine functions. By orthogonality we mean that the integral

$$\begin{aligned}
&\int_0^L \sin\left(\frac{n\pi x}{L}\right) \sin\left(\frac{m\pi x}{L}\right) dx \\
= &\int_0^L \frac{1}{2}\left\{\cos\left[\frac{\pi(n-m)x}{L}\right] - \cos\left[\frac{\pi(n+m)x}{L}\right]\right\} dx \\
= &\frac{L}{2\pi(n-m)} \sin\left[\frac{\pi(n-m)x}{L}\right]\Bigg|_0^L - \frac{L}{2\pi(n+m)} \sin\left[\frac{\pi(n+m)x}{L}\right]\Bigg|_0^L = 0
\end{aligned}$$

for positive integers $n \neq m$. On the other hand, if $n = m$, then

$$\int_0^L \sin\left(\frac{m\pi x}{L}\right) \sin\left(\frac{m\pi x}{L}\right) = \frac{1}{2}\int_0^L \left[1 - \cos\left(\frac{2m\pi x}{L}\right)\right] = \frac{L}{2}$$

for any integer m. The simple way to write this result is

$$\int_0^L \sin\left(\frac{n\pi x}{L}\right) \sin\left(\frac{m\pi x}{L}\right) dx = \frac{L}{2}\delta_{nm}$$

where δ_{nm} is the Kronecker-δ, defined in Section 4.1.2.1.

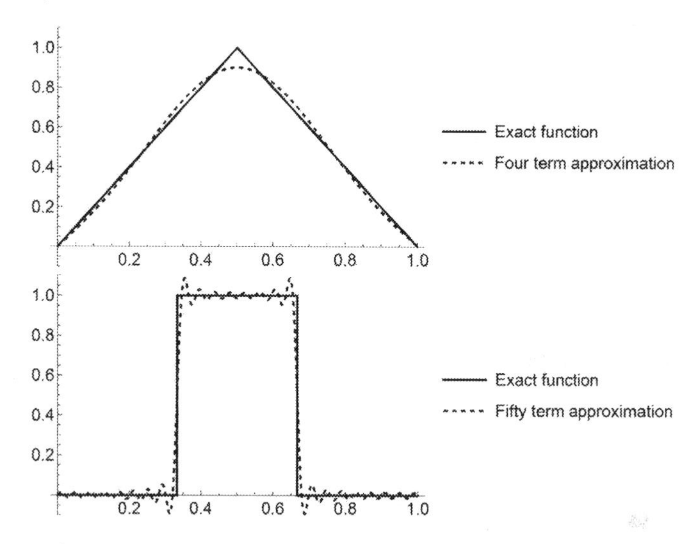

Figure 5.4 Fourier Sine decompositions of two waveforms. The triangle uses only the first four terms in the Fourier series, two of which are in fact zero. Even with so few terms, this approximation is reasonably good. The square wave, however, uses the first 50 terms, and there are still obvious discrepancies at the discontinuities.

Applying this to (5.10) gives

$$\int_0^L f(x)\sin\left(\frac{m\pi x}{L}\right)dx = \sum_{n=1}^{\infty} B_n \int_0^L \sin\left(\frac{n\pi x}{L}\right)\sin\left(\frac{m\pi x}{L}\right)dx = \sum_{n=1}^{\infty} B_n \frac{L}{2}\delta_{nm} = B_m \frac{L}{2}$$

In other words, switching back to the index n,

$$B_n = \frac{2}{L}\int_0^L f(x)\sin\left(\frac{n\pi x}{L}\right)dx \tag{5.11}$$

These are the coefficients of the Fourier Sine Series (5.10). Figure 5.4 shows two examples of Fourier Sine decompositions.

We can now return to our problem of finding the motion of standing waves on a string. Use (5.8) to write the solution for the motion of a string that is fixed at $x = 0$ and $x = L$ and which starts from rest with shape $u(x,0) = f(x)$ as

$$u(x,t) = \sum_{n=1}^{\infty} B_n \sin\left(\frac{n\pi x}{L}\right)\cos\left(\frac{n\pi vt}{L}\right) \tag{5.12}$$

where the coefficients B_n are given by (5.11). It is obvious that this has the shape $f(x)$ at $t = 0$ from our construction of the Fourier Sine Series, but notice also that the different n components oscillate with different frequencies $nv/2L$, so the shape will change as a function of time, as expected.

Figure 5.5 shows the motion of a lopsided triangle wave, using a large number of terms for the Fourier expansion. It is not obvious that this is the behavior you might expect, that is, the trapezoidal motion that maintains the skewness of the initial shape.

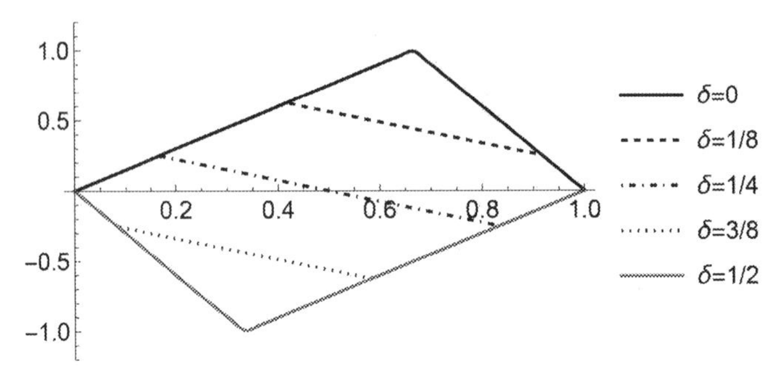

Figure 5.5 Fourier decomposition of a lopsided triangle, using the first 50 terms, showing its motion over time. The decomposition is almost indistinguishable from the exact triangle. The motion is traced out for $t = \delta(2\pi/\omega)$ where ω is the frequency of the lowest frequency component ($n = 1$) and $\delta = 0, 1/8, 1/4, 3/8, 1/2$.

5.3 FOURIER SERIES

As the story goes, Joseph Fourier developed the idea of expanding a function in terms of sines and cosines in order to solve the heat conduction equation in Section 4.5.1.3. As we just saw in Section 5.2, it is also a path for solving the problem of standing waves on a string. These are important hints that this is a generally useful technique.

It apparently took some time for Fourier's ideas to be accepted. The notion that an arbitrary function can be expanded in terms of sines and cosines is not clearly true, and there is some interesting mathematics that goes along with it. I think mathematicians refer to the need to prove "completeness", but we're not going to get into the details. Our point of view will be that if we can find a way to calculate the expansion coefficients, then our job is done.

5.3.1 EXPANSION IN TERMS OF HARMONICS

Our goal is to generalize the Fourier sine series (5.10) to, effectively, include cosine terms. We will also allow the functions we are expanding to be complex, so we write

$$f(x) = \sum_{n=1}^{\infty} \left[A_n e^{ik_n x} + B_n e^{-ik_n x} \right]$$

which we'll call an *expansion in terms of harmonics*. In order to more easily accommodate expressions where $f(x)$ is not necessarily zero anywhere, we will take

$$-\frac{a}{2} \leq x \leq \frac{a}{2} \tag{5.13}$$

for some real, positive value a, which is not necessarily the length L of some string.

Before plowing ahead, let's make some observations of, and changes in, the harmonic expansion above. Firstly, we have excluded $n = 0$ because that term contributes a constant to $f(x)$, and the standing wave boundary conditions implied that constant had to be zero. If we want to consider more general forms of $f(x)$, then we can relax that requirement and include $n = 0$ in the sum.

Note also that the terms for A_n and B_n just flip signs in the exponent, so if we let n be negative, then we only have to do one of the sums. So, now the harmonic expansion is written as

$$f(x) = \sum_{n=-\infty}^{\infty} A_n e^{ik_n x}$$

We will also generalize the boundary conditions. Instead of enforcing $f(x) = 0$ at $x = \pm a/2$, let's use "periodic boundary conditions." That is, the initial shape $f(x)$ isn't constrained to be between $\pm a/2$, but the shape does repeat periodically with period a. Mathematically, this means

$$f(x+a) = f(x)$$

and this gives us a slightly different expression for k_n, namely

$$e^{ik(x+a)} = e^{ikx} \qquad \text{so} \qquad e^{ika} = 1 \qquad \text{or} \qquad ka = 2\pi n$$

for some integer n. Therefore $k = k_n = n2\pi/a$ and we write our general mathematics problem as

$$f(x) = \sum_{n=-\infty}^{\infty} A_n e^{2in\pi x/a} \tag{5.14}$$

where the goal is to find an expression for A_n given an arbitrary $f(x)$.

We get a strong hint on how to find the A_n by integrating both sides over x. Integrating over one period using the region $-a/2 \leq x \leq a/2$,

$$\begin{aligned}
\int_{-a/2}^{a/2} f(x)\,dx &= \sum_{n=-\infty}^{\infty} A_n \int_{-a/2}^{a/2} e^{2in\pi x/a}\,dx \\
&= \sum_{n=-\infty}^{\infty} A_n \frac{a}{2in\pi}\left[e^{in\pi} - e^{-in\pi}\right] = \sum_{n=-\infty}^{\infty} A_n \frac{a}{n\pi}\sin(n\pi)
\end{aligned}$$

Now $\sin(n\pi) = 0$ for any integer n, so it looks like every term on the right side is zero. However, we have to be careful about the $n = 0$ term. In this case

$$\int_{-a/2}^{a/2} f(x)\,dx = A_0 \int_{-a/2}^{a/2} (1)\,dx = A_0 a \qquad \text{or} \qquad A_0 = \frac{1}{a}\int_{-a/2}^{a/2} f(x)\,dx$$

and we see that A_0 is just the average value of the function over one period. This makes good sense. If $n \neq 0$, then the exponential terms are just sines and cosines, all of which integrate to zero over one period.

This tells us how to find the other A_n. Multiply $f(x)$ by $e^{-2im\pi x/a}$ before integrating, where m is some integer. This gives us

$$\begin{aligned}
\int_{-a/2}^{a/2} e^{-2im\pi x/a} f(x)\,dx &= \sum_{n=-\infty}^{\infty} A_n \int_{-a/2}^{a/2} e^{i(n-m)2\pi x/a}\,dx \\
&= \sum_{n=-\infty}^{\infty} A_n \frac{a}{(n-m)\pi}\sin[(n-m)\pi]
\end{aligned}$$

and we have the same situation we had for $m = 0$. That is, every term in the expansion is zero except when $n = m$, in which case

$$\int_{-aL/2}^{a/2} e^{-2im\pi x/a} f(x)\, dx = A_m a$$

Of course, m is just a dummy index, so we have the general expression

$$A_n = \frac{1}{a} \int_{-a/2}^{a/2} e^{-2in\pi x/a} f(x)\, dx \tag{5.15}$$

which is good for all integers n. As far as we're concerned, this confirms that the harmonic expansion (5.14) is valid.

5.3.2 EXAMPLES

Let's try a simple example, namely $f(x) = \sin(2\pi x/a)$, to check the formalism. Note that this $f(x)$ has the correct period, namely $f(x+a) = f(x)$. Write

$$\sin\left(\frac{2\pi x}{a}\right) = \frac{1}{2i} e^{2i\pi x/a} - \frac{1}{2i} e^{-2i\pi x/a}$$

and use (5.15) to get the coefficients

$$A_n = \frac{1}{2ia} \left[\int_{-a/2}^{a/2} e^{2i\pi(1-n)x/a}\, dx - \int_{-a/2}^{a/2} e^{2i\pi(-1-n)x/a}\, dx \right]$$

where both integrals are zero unless $n = \pm 1$. In those cases, we have

$$A_1 = \frac{1}{2ia} \int_{-a/2}^{a/2} (1)\, dx = \frac{1}{2i} \qquad \text{and} \qquad A_{-1} = -\frac{1}{2ia} \int_{-a/2}^{a/2} (1) = -\frac{1}{2i}$$

There are only two terms, therefore, in the expansion (5.14) and we have

$$f(x) = \frac{1}{2i} e^{2i\pi x/a} - \frac{1}{2i} e^{-2i\pi x/a} = \sin\left(\frac{2\pi x}{a}\right)$$

which is of course correct.

The "sawtooth" function $f(x) = 2x/a$ for $-a/2 \leq x \leq a/2$ and which repeats with period a, is a more interesting example. Another is the "square wave," namely

$$f(x) = 1 \text{ for } -\frac{a}{4} \leq x \leq \frac{a}{4} \text{ and } f(x) = -1 \text{ for } --\frac{a}{2} \leq x \leq -\frac{a}{4} \text{ and } \frac{a}{4} \leq x \leq \frac{a}{2}$$

and repeating with period a. These are shown in Figure 5.6, along with their Fourier Series approximations, which are left as exercises.

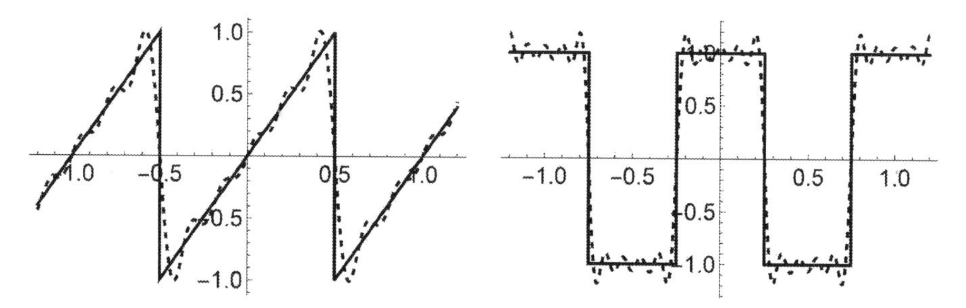

Figure 5.6 Examples of two periodic functions and their Fourier series approximations. The approximations are truncated series based on (5.14), where the coefficients A_n are given by (5.15). The sawtooth function on the left is approximated with terms from $n=-5$ to $n=+5$, whereas the square wave function on the right uses $n=-10$ to $n=+10$.

5.4 PARSEVAL'S THEOREM

Given a function $f(x)$ with period a, that is $f(x+a)=f(x)$ there is a very important theorem that we can prove about the average value of $|f(x)|^2$, namely

$$\frac{1}{a}\int_{-a/2}^{a/2}|f(x)|^2dx=\sum_{n=-\infty}^{\infty}|A_n|^2 \tag{5.16}$$

where the A_n are the coefficients of the Fourier series expansion (5.14). In fact, this theorem is quite easy to prove. All you need to do is insert the expansion twice and then follow your nose. You get

$$\begin{aligned}
\frac{1}{a}\int_{-a/2}^{a/2}|f(x)|^2dx &= \frac{1}{a}\int_{-a/2}^{a/2}f^*(x)\,f(x)\,dx \\
&= \frac{1}{a}\int_{-a/2}^{a/2}\sum_{n=-\infty}^{\infty}A_n^*e^{-2in\pi x/a}\sum_{m=-\infty}^{\infty}A_m e^{2im\pi x/a} \\
&= \frac{1}{a}\sum_{n=-\infty}^{\infty}\sum_{m=-\infty}^{\infty}A_n^*A_m\int_{-a/2}^{a/2}e^{2i(m-n)\pi x/a}dx \\
&= \frac{1}{a}\sum_{n=-\infty}^{\infty}\sum_{m=-\infty}^{\infty}A_n^*A_m\,a\,\delta_{nm}=\sum_{n=-\infty}^{\infty}|A_n|^2
\end{aligned}$$

where the last step recognizes that the double sum collapses to a single sum because all terms are zero except for those with $m=n$.

This results is an example of completeness,, and it has implications everywhere from signal processing to quantum mechanics. We won't be using this further in this course, but you are likely to encounter it later on in your studies.

5.5 FOURIER TRANSFORM

We can now ask ourselves a nice little question. For a periodic function $f(x)$, what happens if the periodicity a becomes infinite? You might suspect this can be useful for a "pulse" that is isolated in space and time. Let's start with the general Fourier series, which we will now write as

$$f(x) = \sum_{n=-\infty}^{\infty} A_n e^{ikx} \tag{5.17}$$

where $k = 2n\pi/a$ and the A_n are given by (5.15).

With $a \to \infty$, and n ranging over all the integers, k will become a continuous variable. Realizing that $\Delta n = 1$, we can write

$$\begin{aligned} f(x) &= \sum_{n=-\infty}^{\infty} A_n e^{ikx} \Delta n = \sum_{n=-\infty}^{\infty} (aA_n) e^{ikx} \Delta\left(\frac{n}{a}\right) = \frac{1}{2\pi} \sum_{n=-\infty}^{\infty} (aA_n) e^{ikx} \Delta k \\ &\underset{a\to\infty}{\longrightarrow} \frac{1}{2\pi} \int_{-\infty}^{\infty} \left[\lim_{a\to\infty} (aA_n)\right] e^{ikx}\, dk \end{aligned}$$

Now, from (5.15), we can write that

$$\lim_{a\to\infty} (aA_n) = \lim_{a\to\infty} \int_{-a/2}^{a/2} e^{-ikx} f(x)\, dx = \int_{-\infty}^{\infty} e^{-ikx} f(x)\, dx$$

We redefine aA_n in the $a \to \infty$ limit to be the *Fourier Transform* of $f(x)$, that is

$$A(k) = \int_{-\infty}^{\infty} e^{-ikx} f(x)\, dx \tag{5.18}$$

The *Inverse Fourier Transform* goes the other way, giving $f(x)$ from $A(k)$, that is

$$f(x) = \frac{1}{2\pi} \int_{-\infty}^{\infty} e^{ikx} A(k)\, dk \tag{5.19}$$

There are different conventions on how to write the Fourier Transform and its inverse. Oftentimes, the signs of k are switched between the two (which pretty much amounts to switching x for k which you're of course free to do.) Sometimes the factor of $1/2\pi$ is "split," putting a factor of $1/\sqrt{2\pi}$ in front of both the transform and the inverse. Another variant is to include a factor of 2π in the exponent inside the integral.

We will use (5.18) and (5.19) as they are, but you should be aware that there is not a universal convention.

5.5.1 THE WIDTH OF THE FOURIER TRANSFORM

The Fourier Transform will find most of its use when we consider "pulses," waveforms that are localized in space. Let's try to come up with a way to characterize the "width" of a localized function.

Consider a function $p(t)$, centered at $t = 0$.[2] We can define a width using a definition inspired by the *standard deviation* from data analysis. (See Section 9.4.1.)

[2]What I really mean is a function that gives an average of zero for its argument. I don't want to spend time being too formal about this. It's of course possible to just translate the argument so that its average value is zero, but in this course we will only deal with functions that are symmetric about zero.

First recognize that for most applications in physics, for example waves in classical mechanics, electromagnetism, and quantum mechanics, the "intensity" is represented by the (absolute value) squared. Therefore, it becomes natural to define the width Δt as

$$(\Delta t)^2 = \frac{1}{\mathscr{I}} \int_{-\infty}^{\infty} t^2 |f(t)|^2 \, dt \qquad \text{where} \qquad \mathscr{I} = \int_{-\infty}^{\infty} |f(t)|^2 \, dt \tag{5.20}$$

is the "integrated intensity." Oftentimes, the pulse shape is normalized so that $\mathscr{I} = 1$.

This definition can be used to find the width Δx of a localized function $f(x)$, as well as the width Δk of its Fourier transform. When we go through some examples of Fourier Transforms, you will see that if $f(x)$ is localized then so is its Fourier Transform $A(k)$. In fact, the "narrower" the pulse $f(x)$, the "wider" will be $A(k)$.

In fact, as we will see shortly, it is not always possible to calculate the width of the Fourier Transform. For many simple forms, the integrals do not converge. We can nevertheless get a "geometric" interpretation of the widths from the shape of the transform.

5.5.2 EXAMPLES OF FOURIER TRANSFORMS

Let's first consider a simple square pulse. That is $f(x) = 1$ for $-a/2 \leq x \leq a/2$, and $f(x) = 0$ otherwise. It is simple to use (5.18) to calculate the Fourier Transform. We have

$$A(k) = \int_{-\infty}^{\infty} e^{-ikx} f(x) \, dx = \int_{-a/2}^{a/2} e^{-ikx} \, dx = \frac{1}{-ik} \left[e^{-ika/2} - e^{ika/2} \right] = \frac{2}{k} \sin\left(\frac{ka}{2}\right)$$

The function and transform are plotted in Figure 5.7. Similar to the way we saw the Fourier Sine Series behave, there are strong oscillations when the function has sharp discontinuities. Notice also that both pulses are localized, the square pulse more so than its transform.

We can calculate the width of $f(x)$ using (5.20). The result is $\Delta x = a/2\sqrt{3}$. However, we cannot apply (5.20) to $A(k)$ in this case, because the integral over all k of $k^2 A(k)$ clearly does not exist. We can, however, take the width of $A(k)$ to be the distance between the two zero crossings on either side of $k = 0$. This gives $\Delta k = 2 \times 2\pi/a = 4\pi/a$. Thus $\Delta x \Delta k = 2\pi/\sqrt{3}$, which is independent of a.

Figure 5.7 also shows the shape and transform of a triangular pulse that goes to zero at $x = \pm a/2$. It is not hard to show that the Fourier Transform for this shape is $(8/ak^2)\sin^2(ak/4)$, which falls to zero more rapidly than for the square pulse. Once again, geometrically, we take $\Delta x = a$ and $\Delta k = 2 \times \pi/4a = \pi/2a$ so that $\Delta x \Delta k = \pi/2$. It seems that this pulse is "narrower" than the square pulse because $\Delta x \Delta k$ is smaller.

As an exercise, you can work out the Fourier Transform of a Gaussian pulse. When you calculate the widths, you will find that $\Delta x \Delta k$ is smaller still.

5.6 THE DIRAC δ-FUNCTION

Here's an interesting question: "What is the Fourier Transform of $f(x) = 1$?" This function is infinitely broad, so do we expect the Fourier Transform to be infinitely narrow? How would we quantify this?

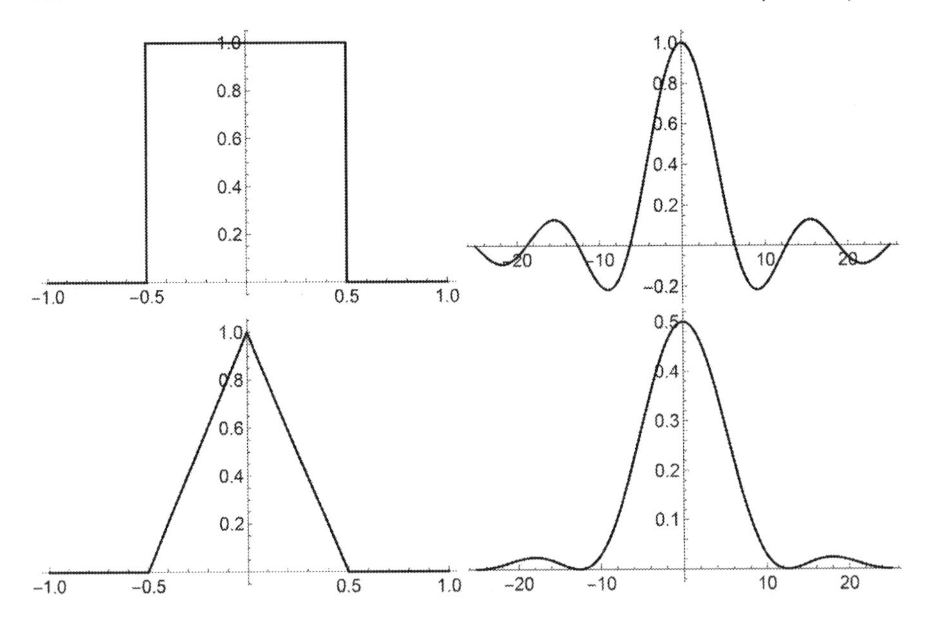

Figure 5.7 Examples of pulses and their Fourier Transforms. The top row shows a square pulse and the bottom shows a triangular pulse. Both the square and triangular pulses become zero at $x = \pm a$. The pulses are plotted versus x/a and the transforms versus ka.

Following (5.18) for $f(x) = 1$ leads us to

$$A(k) = \int_{-\infty}^{\infty} e^{-ikx}\,dx = \int_{-\infty}^{\infty} [\cos(kx) - i\sin(kx)]\,dx$$

The cosines and sines both give zero over any one period, so the integral is zero except for $k = 0$. In that case, we integrate unity over infinity and get infinity. That is, $A(k) = 0$ for $k \neq 0$ but is infinite when $k = 0$. That does sound like something that is infinitely narrow.

We can get more information by considering the integral of $A(k)$ about $k = 0$. Do this by integrating from $-\varepsilon$ to $+\varepsilon$ for some $\varepsilon > 0$. Then

$$\begin{aligned}\int_{-\varepsilon}^{\varepsilon} A(k)dk &= \int_{-\varepsilon}^{\varepsilon} dk \int_{-\infty}^{\infty} e^{-ikx}\,dx = \int_{-\infty}^{\infty} dx \int_{-\varepsilon}^{\varepsilon} dk\, e^{-ikx} = \int_{-\infty}^{\infty} dx \frac{e^{-i\varepsilon x} - e^{i\varepsilon x}}{-ix} \\ &= 2\int_{-\infty}^{\infty} \frac{\sin(\varepsilon x)}{x}\,dx = 2\int_{-\infty}^{\infty} \frac{\sin(y)}{y}\,dy = 4\int_{0}^{\infty} \frac{\sin(x)}{x}\,dx \qquad (5.21)\end{aligned}$$

where I made the substitution $y = \varepsilon x$ in the second-to-last step, and finally recognized that the integral was symmetric about zero and switched the integration variable back to x.

The final integral in (5.21) can in fact be evaluated in several different ways. Here's one way, using a neat trick that seems to be attributed to the physicist Richard

Feynman. The integral we need can be written as

$$I(0)=\int_0^\infty \frac{\sin(x)}{x}\,dx \qquad \text{where} \qquad I(s)=\int_0^\infty e^{-sx}\frac{\sin(x)}{x}\,dx$$

We don't know how to evaluate $I(s)$, but we can evaluate

$$\begin{aligned} I'(s) &= -\int_0^\infty e^{-sx}\sin(x)\,dx = -\frac{1}{2i}\int_0^\infty \left[e^{-sx+ix}-e^{-sx-ix}\right]dx \\ &= -\frac{1}{2i}\left[\frac{e^{-sx+ix}}{-s+i}-\frac{e^{-sx-ix}}{-s-i}\right]_0^\infty = \frac{1}{2i}\left[\frac{1}{-s+i}-\frac{1}{-s-i}\right] = -\frac{1}{1+s^2} \end{aligned}$$

Clearly $I(s)\to 0$ as $s\to\infty$, so

$$I(0) = -\int_0^\infty I'(s)\,ds = \int_0^\infty \frac{1}{1+s^2}\,ds = \tan^{-1}(s)\big|_0^\infty = \frac{\pi}{2}$$

Therefore (5.21) becomes

$$\int_{-\varepsilon}^{\varepsilon} A(k)dk = 2\pi$$

Remarkably, we have been able to quantify the "infinity" at $k=0$ for Fourier Transform of unity. Indeed, we write

$$\delta(k) = \frac{1}{2\pi}\int_{-\infty}^{\infty} e^{-ikx}\,dx \tag{5.22}$$

in which case the Fourier Transform of unity is $A(k)=2\pi\delta(k)$. The function $\delta(k)$ is called the *Dirac δ-function.* You will encounter this (very peculiar) function over and over in the course of studying physics.

Equation (5.22) is only one of very many representations of the Dirac δ-function. For our purposes, *it is sufficient to define $\delta(x)$ as a function which is zero for all $x\neq 0$, but it is "large enough" at $x=0$ so that*

$$\int_{-\varepsilon}^{\varepsilon} \delta(x)\,dx = 1$$

for some $\varepsilon>0$. Some other representations for $\delta(x)$ might include a "box" of width a centered at $x=0$ and with height $1/a$, or a Gaussian function normalized to have unit area in the limit of its width going to zero. The form (5.22) will be particularly useful in quantum mechanics.

A very useful property of the δ-function stems from the relation

$$\int_{-\varepsilon}^{\varepsilon} f(x)\delta(x)\,dx = f(0) \tag{5.23}$$

for some function $f(x)$. This is easy to see. Since ε can be taken as small as we want, it is essentially zero and we can take the $f(x=0)$ out of the integral, leaving us only with the integral of $\delta(x)$.

There are also two-dimensional and three-dimensional versions of the Dirac δ-function, denoted as $\delta^{(2)}(\vec{r})$ and $\delta^{(3)}(\vec{r})$. It is straightforward to write these in Cartesian coordinates, for example $\delta^{(2)}(\vec{r}) = \delta(x)\delta(y)$, but more complicated in polar coordinates. However, it is simplest just to think of these in terms of the fundamental definition of the δ-function, that is $\delta^{(2)}(\vec{r})$ and $\delta^{(3)}(\vec{r})$ are zero for all $\vec{r}$ away from the origin, but

$$\int_S \delta^{(2)}(\vec{r})\, dS = 1 \qquad \text{and} \qquad \int_V \delta^{(3)}(\vec{r})\, dV = 1$$

for any surface S or volume V that encloses the origin.

5.6.1 SURFACE THEOREMS REVISITED

We are now finally ready to resolve the apparent paradox we saw in Section 4.4. There, we saw that the fields $\vec{B}(\vec{r})$ and $\vec{E}(\vec{r})$ given by (4.34) and (4.35) had zero curl and divergence, respectively, but the surface integrals in each case were nonzero. This seemed like a violation of Stokes' Theorem and Gauss' Theorem.

It is clear now, however, that there is one point, namely the origin, for which (4.34) and (4.35) are not really defined because the denominator goes to zero. Integrating over a surface or volume that encloses the origin gives a nonzero result because the curl and divergence are δ-functions at the origin. Indeed, for (4.34), we have

$$\vec{\nabla} \times \vec{B} = 2\pi a\, \delta^{(2)}(\vec{r})\, \hat{k}$$

and for (4.35), we have

$$\vec{\nabla} \cdot \vec{E} = 4\pi a\, \delta^{(3)}(\vec{r})$$

In electromagnetism, $\vec{B}$ is the magnetic field from a long, straight, infinitely thin wire in the z-direction, and $\vec{E}$ is the electric field from a point charge.

EXERCISES

5.1 *Reproduce Figure 5.6.*

5.2 *A function $u(x,t)$ satisfies the wave equation in one dimension x with velocity v. The initial conditions are $u(x,0) = p(x)$, for an arbitrary function $p(x)$, and $\dot{u}(x,0) = 0$. Show that the time development of the wave corresponds to the "splitting" of $p(x)$ into two pieces, one moving to the left and the other moving to the right, each being an exact copy of $p(x)$ but divided by two.*

5.3 *A wave $u(x,t) = g(x+vt)$ moves to the left along a string on the positive x-axis.*

(a) Assume the string is fixed at $x = 0$ so that it cannot move, that is $u(0,t) = 0$. Find the motion of the string for $x \geq 0$ for all times.

(b) Now assume the string is free to move up and down at $x = 0$, and does so in a way that it is always horizontal, that is $\partial u(x,t)/\partial x|_{x=0} = 0$. Once again find the motion of the string for $x \geq 0$ for all times.

5.4 *Use a Fourier Sine decomposition to find the motion of a string that is fixed at $x = 0$ and $x = L$, and that starts from rest with an initial shape $u(x,0) = (2/L)^4 x^2 (x - L)^2$. I encourage you do this problem using a computer and generating an animation.*

5.5 *A string with fixed ends at $x = 0$ and $x = L$ has a mass density μ per unit length and is under tension T. Find the vertical motion $u(x,t)$ of the shape of the string if it is initially flat, that is $u(x,0) = 0$, but has an initial vertical velocity profile $\dot{u}(x,0) = V \sin(3\pi x/L)$.*

5.6 *A string with fixed ends at $x = 0$ and $x = L$ has an initial shape given by*

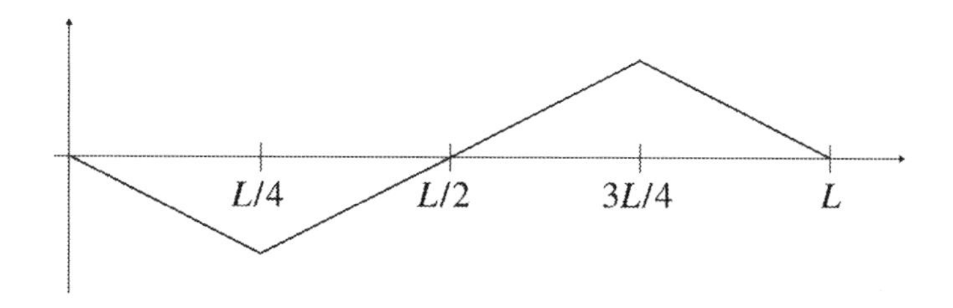

Use a Fourier decomposition to find the shape of the string as a function of time, using enough terms in the expansion so that the true shape is clear. Plot the shape at various times within one fundamental period (or make an animation).

5.7 *A string with fixed ends at $x = 0$ and $x = L$ has a total mass M and is under tension T. Assume the string is vibrating in normal mode n. Integrate over the length of the string to find its total kinetic energy as a function of time. You can assume the solution for the string motion where it is initially at rest, and express your result in terms of the Fourier coefficients B_n in (5.12).*

5.8 *Find the Fourier Transform $A(k)$ of a pulse $f(x) = C(x^2 - a^2)^2$, where C is a constant, for $-a \le x \le a$ and $f(x) = 0$ for $|x| > a$. Plot $f(x)$ and $A(k)$ and briefly compare them. Find the RMS width of $f(x)$ and of $A(k)$, and show that their product is independent of a.*

5.9 *Find the Fourier Transform $A(k)$ of the function $f(x) = ae^{-\beta|x|}$ where $\beta > 0$.*

5.10 *This problem concerns the Fourier Transform and width of the Gaussian function*

$$f(x) = \frac{1}{\sigma\sqrt{2\pi}} e^{-x^2/2\sigma^2}$$

(a) *Find the Fourier Transform $A(k)$ of $f(x)$. The integral is not hard to do. Just complete the square in the exponent, and use what you know about Gaussian integrals.*

(b) *Calculate the "width" Δx (that is, the square root of the variance) of $f(x)$. (See the notes for details.) Again make use of what you know about Gaussian integrals.*

(c) *Next find the width* Δk *of* $A(k)$.
(d) *Determine the product* $\Delta x \Delta k$. *How does this compare to the same result for the triangle pulse that we derived in class?*

5.11 *Use the generalized definition of the* δ*-function to show that*

$$\delta(x) = \lim_{n \to \infty} \frac{n}{\sqrt{\pi}} e^{-n^2 x^2}$$

5.12 *Show that the following relationships are consistent with the fundamental definition of the* δ*-function. You can make use of results derived in the notes, but you'll likely find it useful to employ integration by parts.*

(a) $x\delta(x) = 0$
(b) $x\delta'(x) = -\delta(x)$
(c) $x^2\delta''(x) = 2\delta(x)$

6 Vectors and Matrices

This chapter concerns the subject known as *Linear Algebra*. As with so many things in this course, this is a very large subject so we are only able to scratch the surface.

I will introduce the subject with a very specific example that I can use to set the stage, namely the problem of how to solve systems of linear algebraic equations. From there, I will get more formal and generalize to other kinds of physical systems.

6.1 INTRODUCTION TO SYSTEMS OF LINEAR EQUATIONS

How would you go about solving the system of equations

$$2x+y = 3 \tag{6.1a}$$

$$x-y = 0 \tag{6.1b}$$

for the variables x and y? A glance at the second equation tells you that $x = y$, and in your head you see that this gives $x = y = 1$ from the first equation. Of course, this is a very simple example, but you know that, in principle, you are able to multiply either or both of the equations by constants, add or subtract the equations from each other, and manipulate things one way or another so that you can isolate x and y.

Sometimes you encounter pitfalls, though. Consider solving the system of equations

$$2x+y = 3 \tag{6.2a}$$

$$4x+2y = 6 \tag{6.2b}$$

for x and y. If you multiply the first equation by 2 and subtract it from the second equation, you end up with $0 = 0$, which is true, but useless for finding x and y. (If the right hand side of the second equation was something other than 6, you wouldn't even end up with a true statement.) The problem, of course, is that these two equations are not *independent*. That is, they are really the same equation, because the coefficients of x and y in the second equation are both just the same factor, namely 2, of the coefficients in the first equation.

Our first job in this chapter is to formalize how we will write systems of linear equations and this will lead us into a discussion of *vectors* and *matrices*. The concept of a "vector" will be a very expanded version of what we discussed in Section 4.1. Seeing this formalism will create some obvious questions regarding how we solve systems of linear equations, so we'll next address those questions before coming back to solving these systems in Section 6.3.9.

So let's think about a system of equations with more than just two variables x and y. Rather than run through the alphabet, let's say there are N equations for N variables x_1, x_2, $\ldots x_N$. We'll use an upper case "A" to denote the coefficients with

DOI: 10.1201/9781003355656-6

appropriate subscripts, and write

$$\begin{array}{ccccccccccc} A_{11}x_1 & + & A_{12}x_2 & + & A_{13}x_3 & + & \cdots & + & A_{1N}x_N & = & c_1 \\ A_{21}x_1 & + & A_{22}x_2 & + & A_{23}x_3 & + & \cdots & + & A_{2N}x_N & = & c_2 \\ A_{31}x_1 & + & A_{32}x_2 & + & A_{33}x_3 & + & \cdots & + & A_{3N}x_N & = & c_3 \\ \vdots & & \vdots & & \vdots & & & & \vdots & & \vdots \\ A_{N1}x_1 & + & A_{N2}x_2 & + & A_{N3}x_3 & + & \cdots & + & A_{NN}x_N & = & c_N \end{array} \tag{6.3}$$

where the c_N are just the numbers on the right hand sides of the equations. These equations are the same as writing

$$\sum_{j=1}^{N} A_{ij}x_j = c_i$$

where $i = 1, 2, 3, \ldots, N$. Using our summation notation agreement, where any index repeated twice is automatically summed over, we get the compact form

$$A_{ij}x_j = c_i \tag{6.4}$$

which is exactly equivalent to (6.3).

An even more economical way to write (6.3) or (6.4), which in fact is profound, is

$$\underline{\underline{A}}\,\underline{x} = \underline{c} \tag{6.5}$$

where

$$\underline{\underline{A}} = \begin{bmatrix} A_{11} & A_{12} & A_{13} & \cdots & A_{1N} \\ A_{21} & A_{22} & A_{23} & \cdots & A_{2N} \\ A_{31} & A_{32} & A_{33} & \cdots & A_{3N} \\ \vdots & \vdots & \vdots & & \vdots \\ A_{N1} & A_{N2} & A_{N3} & \cdots & A_{NN} \end{bmatrix} \quad \underline{x} = \begin{bmatrix} x_1 \\ x_2 \\ x_3 \\ \vdots \\ x_N \end{bmatrix} \quad \text{and} \quad \underline{c} = \begin{bmatrix} c_1 \\ c_2 \\ c_3 \\ \vdots \\ c_N \end{bmatrix} \tag{6.6}$$

define the (square) *matrix* $\underline{\underline{A}}$ and the *column vectors* $\underline{x}$ and $\underline{c}$. You can now easily visualize the sum (6.4) by imagining taking the top row of $\underline{\underline{A}}$, matching it against the column $\underline{x}$, multiplying each element by each other one-by-one, adding this up and equating it to the first element of $\underline{c}$, that is c_1. Then repeat for the second row and equate to c_2 and so forth.

Please take note that the notation that I'm using, a double underline for a matrix and a single underline for a column vector, is not anyone's standard so far as I can tell. Different books use different notations, and some simply let you figure out what things are from context. I am using this notation because it is easy to write on the board in class.

Clearly, one can define a matrix that is not square. A proper course in Linear Algebra will go through the properties of these kinds of matrices as well. In physics, however, we rarely encounter matrices that are not square, so I just won't bother to go there.

Notice that our matrix notation (6.5) is easily generalized to products of matrices. That is, something like $\underline{\underline{A}}\,\underline{\underline{B}}$ has elements

$$\left[\underline{\underline{A}}\,\underline{\underline{B}}\right]_{ij} = \left(\sum_{k=1}^{N}\right) A_{ik}B_{kj} \tag{6.7}$$

where I put parenthesis around the summation only to warn you that I'm going to stop writing the summation symbol and resort to our summation convention instead.

It is important to notice that in (6.7), the summed index k appears adjacently in $A_{ik}B_{kj}$. This is a very useful way to remember the order of matrices and vectors when you are writing things explicitly in terms of their elements and indices.

The order in which you write matrices and vectors matters! You can talk about $\underline{\underline{A}}\,\underline{x}$ for an $N \times N$ matrix $\underline{\underline{A}}$ and N-dimensional column vector $\underline{x}$. but $\underline{x}\underline{\underline{A}}$ is nonsense. Also, $\underline{\underline{A}}\,\underline{\underline{B}}$ and $\underline{\underline{B}}\,\underline{\underline{A}}$ are both legal, if $\underline{\underline{A}}$ and $\underline{\underline{B}}$ are both $N \times N$ matrices, but in general $\underline{\underline{A}}\,\underline{\underline{B}} \neq \underline{\underline{B}}\,\underline{\underline{A}}$.

There is a very special and important matrix called the *identity matrix* $\underline{\underline{I}}$. Its elements are all zero except for the diagonal elements, which are all unity. That is

$$\underline{\underline{I}} = \begin{bmatrix} 1 & 0 & 0 & \cdots & 0 \\ 0 & 1 & 0 & \cdots & 0 \\ 0 & 0 & 1 & \cdots & 0 \\ \vdots & \vdots & \vdots & & \vdots \\ 0 & 0 & 0 & \cdots & 1 \end{bmatrix} \qquad \text{i.e.} \qquad \left[\underline{\underline{I}}\right]_{ij} = \delta_{ij} \tag{6.8}$$

(Recall the Kronecker δ symbol from Section 4.1.2.1.) It should be clear that $\underline{\underline{I}}\,\underline{x} = \underline{x}$ for any column vector $\underline{x}$ and $\underline{\underline{I}}\,\underline{\underline{A}} = \underline{\underline{A}} = \underline{\underline{A}}\,\underline{\underline{I}}$ for any matrix $\underline{\underline{A}}$.

There is one more thing to show you about systems of linear equations before we jump off into the formalism of matrices, vectors, vector spaces, and operators.

In principle, for some matrix $\underline{\underline{A}}$ there can exist an *inverse matrix* $\underline{\underline{A}}^{-1}$ such that

$$\underline{\underline{A}}^{-1}\underline{\underline{A}} = \underline{\underline{I}} = \underline{\underline{A}}\underline{\underline{A}}^{-1} \tag{6.9}$$

Armed with the inverse matrix, we can immediately write down the solution for the x_i represented in (6.5) simply by multiplying both sides by $\underline{\underline{A}}^{-1}$. That is

$$\underline{x} = \underline{\underline{A}}^{-1}\underline{c}$$

This gives you a glimpse of the practical power of using matrices to solve the systems of linear equations. The pitfalls I mentioned at the start of this section will correspond to circumstances under which the inverse does not exist for a particular matrix $\underline{\underline{A}}$.

From here, we will talk about what we really mean by a *vector*, namely as an element of a *vector space*. An *operator* can act on a vector to turn it into a different vector. In most cases, we can *represent* a vector by a column vector, and an operator by a matrix, but there are a lot of blanks to fill in before we get into this in any level of detail.

6.2 GENERALIZED VECTORS

Section 4.1.5 gave you a hint that a "vector" is actually much more than a collection of two or three coordinates that specify the position in a plane or in space. In fact, a mathematical definition of a vector requires a good deal of sophistication in order to define a *vector space*. A *vector* is an element of a vector space.

I'm going to take a more practical approach here. For one thing, we don't have a lot of time and I'm afraid of getting some of you lost in the formalism if I tried. (Nevertheless, I urge you to take a math course in Linear Algebra at some point.) Secondly, and more importantly, I think it is better to take an incremental approach to abstract notion of vectors. By now you should be comfortable with the vector as a quantity in two-dimensional or three-dimensional space. The abstraction level I will take you to now will be for an N-dimensional space with vectors that might contain complex numbers.

You will find much more complete descriptions of vector spaces in any one of the many comprehensive textbooks available on mathematical methods in physics.

6.2.1 THE N-DIMENSIONAL COMPLEX VECTOR

This is the entity that we will consider a "generalized vector" in this course. As I said, it is not as general as it could be, but will work for us.

A *vector* $\underline{v}$ is a collection of N (possibly) complex numbers $v_1, v_2, \ldots, v_N$. That is $\underline{v} \in \mathbb{C}^N$. If we want to write $\underline{v}$ in terms of its specific elements, we do so with a *column vector*, namely

$$\underline{v} = \begin{bmatrix} v_1 \\ v_2 \\ \vdots \\ v_N \end{bmatrix}$$

We refer to the individual v_i as *components* of the vector. The collection of all possible vectors forms a *vector space*.

It is also possible to represent a vector by its *transpose*, that is

$$\underline{v}^{\mathsf{T}} = \begin{bmatrix} v_1 & v_2 & \cdots & v_N \end{bmatrix}$$

which we also refer to as a *row vector*. We will in fact more often use the *Hermitian transpose*

$$\tilde{\underline{v}} = \underline{v}^{*\mathsf{T}} = \begin{bmatrix} v_1^* & v_2^* & \cdots & v_N^* \end{bmatrix}$$

which is the transpose of the complex conjugates of the components of $\underline{v}$. Sometimes, especially in Quantum Mechanics, we refer to the space of Hermitian conjugate vectors as the *dual space*.

Multiplying a vector by a (complex) number means to multiply each component by that number, that is

$$c\underline{v} = \begin{bmatrix} cv_1 \\ cv_2 \\ \vdots \\ cv_N \end{bmatrix}$$

for some complex number c. Clearly, $c\underline{v}$ is a member of the vector space, assuming that $\underline{v}$ is a member.

Addition must be a property of a vector space. Two vectors $\underline{u}$ and $\underline{v}$ can be added by adding their components, that is

$$\underline{u} + \underline{v} = \begin{bmatrix} u_1 + v_1 \\ u_2 + v_2 \\ \vdots \\ u_N + v_N \end{bmatrix}$$

and the result is also clearly a member of the vector space. In fact, the actions of addition and multiplication mean that any linear combination of two vectors is a vector. That is

$$\underline{w} = a\underline{u} + b\underline{v}$$

is a member of the same vector space as $\underline{u}$ and $\underline{v}$ where a and b are complex numbers.

The vector space needs to have an *identity* element under addition. This is of course the vector $\underline{0}$ where all components are zero. Furthermore, each element of the vector space needs to have an inverse, that is, something to which it can be added giving the result $\underline{0}$. For a vector $\underline{v}$, the inverse is clearly $-\underline{v} = c\underline{v}$ where $c = -1$.

I remind you again that I am using a rather specific definition of a vector, much more general than the simple object in 2D or 3D real space, but less general than in fact is possible. I'll come back to this point in Section 6.2.5.

6.2.2 INNER PRODUCT AND NORM

The *inner product*[1] $\langle u|v\rangle$ of two vectors $\underline{u}$ and $\underline{v}$ means to take the Hermitian conjugate of the first one and multiply it by the second one, in the sense that you are multiplying a column vector by a row vector as if they were matrices. That is, the inner product is

$$\langle u|v\rangle = \tilde{\underline{u}}\,\underline{v} = u_i^* v_i$$

where I have employed the summation notation. Clearly $\langle u|v\rangle = \langle v|u\rangle^*$.

It should be obvious to you that the inner (i.e. dot) product of two 3D (real) vectors $\vec{a}$ and $\vec{b}$ is completely consistent with this definition. Indeed, if the inner product of two vectors is zero, we say that they are *orthogonal*.

[1] I am borrowing the notation $\langle u|v\rangle$ from quantum mechanics.

The inner product of a vector with itself gives the square of the *norm* of the vector, that is

$$\langle v|v\rangle = v_i^* v_i = \sum_{i=1}^{N} |v_i|^2$$

Don't be surprised if every now and then I slip up and refer to the *length* or *magnitude* of a vector instead of the norm. This is an obvious throwback to the physical notion of a vector measuring the location of some object in three-dimensional space.

6.2.3 UNIT VECTORS

Unit vectors are vectors whose norm equals unity. A particularly useful set of unit vectors are those where one component equals one and all other components equal zero. We might denote these unit vectors as

$$\underline{e}_i = \begin{bmatrix} 0 \\ 0 \\ \vdots \\ 1 \\ \vdots \\ 0 \end{bmatrix}$$

where the "1" is in the ith row of the column vector. It is clear that any vector $\underline{v}$ can be written as a linear combination of the unit vectors. That is

$$\underline{v} = v_i \, \underline{e}_i$$

where, again, we invoke the summation convention. (I'll stop staying this at some point.) We might say something like "the vector space is spanned by the unit vectors" because we can construct any vector using them in this way.

We say that the unit vectors form a *basis* for the vector space.

6.2.4 DYADICS AND TENSORS

How might you think about an object like $\underline{u}\,\tilde{\underline{v}}$, that is the "product" of a vector and its Hermitian conjugate, but in the "wrong" order? It seems reasonable to think about this sort of thing in terms of matrix multiplication, and write

$$\underline{u}\,\tilde{\underline{v}} = \begin{bmatrix} u_1 \\ u_2 \\ \vdots \\ u_N \end{bmatrix} \begin{bmatrix} v_1^* & v_2^* & \cdots & v_N^* \end{bmatrix} = \begin{bmatrix} u_1 v_1^* & u_1 v_2^* & \cdots & u_1 v_N^* \\ u_2 v_1^* & u_2 v_2^* & \cdots & u_2 v_N^* \\ \vdots & \vdots & \cdots & \vdots \\ u_N v_1^* & u_N v_2^* & \cdots & u_N v_N^* \end{bmatrix}$$

In other words, this "wrong" product is a matrix. This actually turns out to be a very useful construction for physics, and you will see it in various courses. It is a good way to understand the moment of inertia of rigid bodies in classical mechanics, for example.

We call this kind of construction a *dyadic* or *dyad*. It is a special form of something called a *tensor*. We won't spend much time on the concept of tensors in this course, but you will encounter them elsewhere in your studies of physics.

6.2.5 FUNCTIONS CAN FORM A VECTOR SPACE

Even though we talked about the complex vector in Section 6.2.1 as our context for this course, I will take a moment to talk about an important generalization beyond that. I'm inspired to do this because of a question a student asked in class about what I meant when I referred to "orthogonal functions."

Recall from Section 5.3.1 that we argued that any function $f(x)$ with periodicity defined by $f(x+a) = f(x)$ could be written as a linear combination of the functions $e^{2in\pi x/a}$ where n is some integer. If we replace the idea of "inner product" of two vectors by the integration of two functions over the relevant domain, that is for any two functions $g(x)$ and $h(x)$ we write

$$\langle g(x)|h(x)\rangle = \int_{-a/2}^{a/2} g^*(x)h(x)\,dx$$

then we have a well-defined vector space. The functions $(1/a)e^{2in\pi x/a}$ where $n \in \mathbb{Z}$ are the unit vectors that form the basis for any function that is a member of the vector space. It is pretty easy to see that they have unit norm, and are orthogonal to each other.

Another, and perhaps more interesting, example concerns the Legendre Polynomials $P_\ell(x)$ from Section 3.6.4. Although we didn't prove it, the $P_\ell(x)$ obey an orthogonality relationship

$$\int_{-1}^{1} P_n(x)P_m(x)\,dx = \frac{2}{2n+1}\delta_{nm}$$

(I did not bother to take the complex conjugate because the Legendre Polynomials are real.) This means that we could argue, in exactly the same way we did for Fourier Series, that any function defined over the range $-1 \le x \le 1$ can be written as a linear combination of the Legendre Polynomials. That is, the (appropriately normalized) $P_\ell(x)$ form the basis vectors for the vector space of polynomials.

You will encounter both of these examples when you study Quantum Mechanics.

6.3 OPERATIONS ON VECTORS: MATRICES

If we were talking about vector spaces in the abstract, the next thing to talk about would be "operations" on vectors. An *operator* is an object which can transform a vector into another vector. Abstract operators are critical in the formulation of Quantum Mechanics, for example.

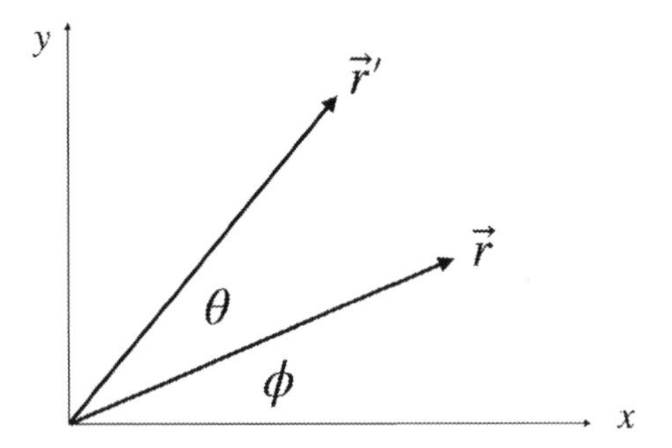

Figure 6.1 Rotation of a vector $\vec{r} = \hat{i}x + \hat{j}y$ by an angle θ into a new vector $\vec{r}' = \hat{i}x' + \hat{j}y'$. In polar coordinates, both vectors have the same magnitude r, but the polar angle of $\vec{r}$ is ϕ, and the polar angle of $\vec{r}'$ is $\phi + \theta$.

In our context, however, we will be representing operators on vectors as matrices. See, for example, (6.5) where the matrix $\underline{\underline{A}}$ operates on a vector $\underline{x}$ and transforms it into a vector $\underline{c}$. This is how we will formalize operations on vectors.

Matrices and their actions on vectors arrive naturally in almost ever area of physics. In fact, we've already seen in Section 3.7.1 how this works in the coupled simple harmonic oscillator, although I didn't really show you that we were working with matrices and vectors. Nevertheless, extending this discussion to N masses, and important problem in classical mechanics known as "N-body oscillations," will rely heavily on matrix and vector formalism.

A common use of operators with which you are already familiar are rotations in ordinary three-dimensional (or two-dimensional) space. A position vector pointed in any direction can be rotated into a vector in a different direction, and that operation is performed by multiplying a vector by a rotation matrix.

Let's illustrate this in two dimensions. Figure 6.1 shows the rotation of $\vec{r} = \hat{i}x + \hat{j}y$ into a vector $\vec{r}' = \hat{i}x' + \hat{j}y'$ through an angle θ. The transformation from x and y into x' and y' is easy enough to figure out. We have

$$\begin{aligned} x' &= r\cos(\phi+\theta) = r\cos\phi\cos\theta - r\sin\phi\sin\theta \\ &= x\cos\theta - y\sin\theta \\ y' &= r\sin(\phi+\theta) = r\sin\phi\cos\theta + r\cos\phi\sin\theta \\ &= x\sin\theta + y\cos\theta \end{aligned}$$

which we can write in our vector and matrix notation now as

$$\underline{r}' = \underline{\underline{R}}\,\underline{r}$$

where

$$\underline{r}' = \begin{bmatrix} x' \\ y' \end{bmatrix} \quad \text{and} \quad \underline{r} = \begin{bmatrix} x \\ y \end{bmatrix}$$

are just the spatial vectors $\vec{r}'$ and $\vec{r}$ written as generalized column vectors, and

$$\underline{\underline{R}} = \begin{bmatrix} \cos\theta & -\sin\theta \\ \sin\theta & \cos\theta \end{bmatrix} \tag{6.10}$$

is what we would call the "rotation matrix."

Furthermore, as I've already mentioned, Quantum Mechanics is formalized in terms of operators and (abstract) vectors, where these are "represented" by matrices and column vectors. Special Relativity is best formulated in terms of "four vectors" which represent "position" in spacetime, that is, a point in 3D space at a particular time, and "translating to a new reference frame" means to operate on a four vector with the matrix of the Lorentz Transformation.

So now let's gather up many of the things we've been saying, and formalize them a bit. Much of what follows will therefore be a repeat of things earlier in this chapter, but I thought it would be a good idea to collect things into the same place. There is a lot of terminology that goes along with matrix algebra, so I'm hoping this section will be a handy reference.

6.3.1 MATRICES MULTIPLYING VECTORS

A matrix $\underline{\underline{A}}$ multiplies a vector $\underline{u}$ creating a new vector $\underline{v}$ as

$$\underline{\underline{A}}\,\underline{u} = \underline{v} \tag{6.11}$$

If the vectors have dimension N, then $\underline{\underline{A}}$ is an $N \times N$ *square matrix*. We write $\underline{\underline{A}}$ in terms of its *elements* as

$$\underline{\underline{A}} = \begin{bmatrix} A_{11} & A_{12} & A_{13} & \cdots & A_{1N} \\ A_{21} & A_{22} & A_{23} & \cdots & A_{2N} \\ A_{31} & A_{32} & A_{33} & \cdots & A_{3N} \\ \vdots & \vdots & \vdots & & \vdots \\ A_{N1} & A_{N2} & A_{N3} & \cdots & A_{NN} \end{bmatrix} \tag{6.12}$$

Equation (6.11) actually represents the N equations

$$A_{ij}u_j = v_i$$

where each value of i is a different equation. In the context of operators, the matrix $\underline{\underline{A}}$ operates on the vector $\underline{u}$ giving the vector $\underline{v}$.

To pictorially understand the operation (6.11) we imagine that the first row of $\underline{\underline{A}}$ is pulled out of the matrix, turned upright to line up alongside the column vector $\underline{u}$, and then each element of the two columns multiplied by each other and added up.[2] This sum then is set equal to the first row of the new vector $\underline{v}$. The process then repeats, row by row, until you've calculated all N elements of $\underline{v}$.

[2]Note how each instance of the repeated index j in this equation is adjacent to the other. Recall my comment on Page 151. We will see this again when we multiply matrices in Section 6.3.2.

This is all perfectly consistent with a different multiplication operation, namely

$$\underline{\tilde{u}}\,\underline{\underline{A}} = \underline{\tilde{v}} \tag{6.13}$$

which is that a row vector can multiply a matrix giving a new row vector. By elements,

$$u_i^* A_{ij} = v_j^*$$

Here, pictorially, the row vector is tilted upright, lined up against the columns of the matrix one-by-one, multiplying each pair and adding them up, repeating for each column. It is fair to think of this as the operator represented by $\underline{\underline{A}}$ operating *to the left* on the row (or dual) vector $\underline{\tilde{u}}$. This concept will be found all through Quantum Mechanics.

Since we now think of $\underline{\underline{A}}\,\underline{v}$ as a new (column) vector, or $\underline{\tilde{u}}\,\underline{\underline{A}}$ as a new (row) vector, the quantity

$$\langle u|A|v\rangle \equiv \underline{\tilde{u}}\,\underline{\underline{A}}\,\underline{v} = u_i^* A_{ij} v_j \tag{6.14}$$

is well-defined by the properties of the inner product. In fact, in Quantum Mechanics we refer to a quantity like this as a *matrix element.*

6.3.2 MATRICES MULTIPLYING OTHER MATRICES

If the equation $\underline{\underline{A}}\,\underline{u} = \underline{w}$ means that $\underline{w}$ is created by operating on $\underline{u}$ with $\underline{\underline{A}}$, but $\underline{u}$ was in fact created by the action of $\underline{\underline{B}}$ on a different vector $\underline{v}$, that is $\underline{u} = \underline{\underline{B}}\,\underline{v}$, then we would write

$$\underline{\underline{A}}\,\underline{\underline{B}}\,\underline{v} = \underline{w}$$

which implies that we can create a new matrix $\underline{\underline{C}} = \underline{\underline{A}}\,\underline{\underline{B}}$ with elements

$$C_{ij} = A_{ik} B_{kj}$$

where we note the placement of the inner indices k and the outer indices i and j. Pictorially, this means we pick out the first row of $\underline{\underline{A}}$, tilt it upright and line it up against the first column of $\underline{\underline{B}}$, multiply the pairs and sum up the results, and this becomes the first element of $\underline{\underline{C}}$ in the upper left corner. Then repeat with the first row of $\underline{\underline{A}}$ and the *second* column of $\underline{\underline{B}}$ to get the second element in the first row of $\underline{\underline{C}}$. Repeat this for all the columns of $\underline{\underline{B}}$, and then again for the second row of $\underline{\underline{A}}$, and so forth.

There is no reason, a priori, that the vector you get from multiplying $\underline{\underline{B}}$ on $\underline{v}$ and then multiplying the result by $\underline{\underline{A}}$ should be the same as if you first multiply $\underline{v}$ by $\underline{\underline{A}}$ and then multiply the result by $\underline{\underline{B}}$. In other words, in general

$$\underline{\underline{A}}\,\underline{\underline{B}} \neq \underline{\underline{B}}\,\underline{\underline{A}}$$

We say that *matrix multiplication is not commutative*. There is a lot of physics in knowing whether or not certain matrices commute.

6.3.3 SYMMETRIC AND DIAGONAL MATRICES

A matrix is called *symmetric* if the elements above the diagonal are equal, pairwise, to the elements below the diagonal, just by flipping indices. That is, a matrix $\underline{\underline{A}}$ is symmetric if its elements obey

$$A_{ij} = A_{ji}$$

We will see a cleaner definition of a symmetric matrix in Section 6.3.4.

Certain matrices are called *diagonal matrices* if the only nonzero elements are along the diagonal. For example

$$\underline{\underline{A}} = \begin{bmatrix} a^{(1)} & 0 & \cdots & 0 \\ 0 & a^{(2)} & \cdots & 0 \\ \vdots & \vdots & \cdots & \vdots \\ 0 & 0 & \cdots & a^{(N)} \end{bmatrix}$$

Or, in term of elements,

$$A_{ij} = a^{(i)}\delta_{ij} = a^{(j)}\delta_{ij}$$

where we are <u>not</u> invoking the summation notation because the first i (or j) is a superscript in parentheses. An important special case is the identity matrix $\underline{\underline{I}}$ with elements $I_{ij} = \delta_{ij}$.

Clearly, all diagonal matrices are symmetric, since $\delta_{ij} = \delta_{ji}$.

It is worth noting that all diagonal matrices commute with each other, and their product is also diagonal, with the diagonal elements of the product matrix just equal to the individual element products. That is, if $\underline{\underline{A}}$ and $\underline{\underline{B}}$ are both diagonal, then

$$\left(\underline{\underline{A}}\,\underline{\underline{B}}\right)_{ij} = a^{(i)}\delta_{ik}b^{(j)}\delta_{kj} = a^{(i)}b^{(j)}\delta_{ij} = b^{(i)}a^{(j)}\delta_{ij} = b^{(i)}\delta_{ik}a^{(j)}\delta_{kj} = \left(\underline{\underline{B}}\,\underline{\underline{A}}\right)_{ij}$$

6.3.4 TRANSPOSE AND HERMITIAN TRANSPOSE OF A MATRIX

The *transpose* $\underline{\underline{A}}^{\mathsf{T}}$ of a matrix $\underline{\underline{A}}$ has the same components as the original matrix but with its rows and columns reversed. That is, in terms of elements

$$\left(\underline{\underline{A}}^{\mathsf{T}}\right)_{ij} = \left(\underline{\underline{A}}\right)_{ji}$$

Clearly, a succinct definition of a symmetric matrix is just when a matrix equals its transpose. In other words, $\underline{\underline{A}}$ is a symmetric matrix if

$$\underline{\underline{A}}^{\mathsf{T}} = \underline{\underline{A}}$$

Given that our vectors and matrices are possibly made of complex numbers, we will find it handy to also define the *Hermitian transpose* which is the transpose of the complex conjugates of the elements of the original matrix, completely analogous to the situation with vectors. That is the Hermitian transpose[3] $\tilde{\underline{\underline{A}}}$ of a matrix $\underline{\underline{A}}$ has elements

$$\left(\tilde{\underline{\underline{A}}}\right)_{ij} = \left(\underline{\underline{A}}\right)^*_{ji}$$

If the Hermitian transpose leaves the matrix unchanged, we say the matrix is *Hermitian*. That is, $\underline{\underline{A}}$ is Hermitian if

$$\tilde{\underline{\underline{A}}} = \underline{\underline{A}}$$

It is easy to show that the transpose of a product of matrices is the product of the transposed matrices, but in the reverse order. That is

$$\left(\underline{\underline{A}}\,\underline{\underline{B}}\right)^{\mathsf{T}} = \underline{\underline{B}}^{\mathsf{T}}\underline{\underline{A}}^{\mathsf{T}} \tag{6.15}$$

Just write this out in terms of elements to prove the assertion. We have

$$\left[\left(\underline{\underline{A}}\,\underline{\underline{B}}\right)^{\mathsf{T}}\right]_{ij} = \left(\underline{\underline{A}}\,\underline{\underline{B}}\right)_{ji} = A_{jk}B_{ki} = \left(\underline{\underline{A}}^{\mathsf{T}}\right)_{kj}\left(\underline{\underline{B}}^{\mathsf{T}}\right)_{ik} = \left(\underline{\underline{B}}^{\mathsf{T}}\right)_{ik}\left(\underline{\underline{A}}^{\mathsf{T}}\right)_{kj} = \left(\underline{\underline{B}}^{\mathsf{T}}\underline{\underline{A}}^{\mathsf{T}}\right)_{ij}$$

Notice how I made use of the "adjacent indices" association with the matrix product by putting the two k indices next to each other in the second-to-last step.

It is straightforward to prove that the same thing holds for the Hermitian transpose. That is

$$\widetilde{\underline{\underline{A}}\,\underline{\underline{B}}} = \tilde{\underline{\underline{B}}}\tilde{\underline{\underline{A}}} \tag{6.16}$$

6.3.5 DETERMINANT OF A MATRIX

The *determinant* $|\underline{\underline{A}}|$ of a matrix $\underline{\underline{A}}$ is an extremely important concept which unfortunately is difficult to clearly define.[4] Happily, we don't have to calculate the determinant often, and can generally leave that task to some computer application. I will nevertheless go through the basics here.

The determinant is a peculiar thing, mathematically. It maps the matrices, which we'd write as the Cartesian product $\mathbb{C}^N \times \mathbb{C}^N$ onto $\mathbb{C}$. That is, it takes a very large set and maps it into a much smaller set. (Never mind that both sets are actually infinite.)

Probably the best way to think about the determinant is as the sum of the terms formed from every possible product of the elements of the matrix, picking from each row and one from each column, but never repeating the row or column, and including an alternating sign. You could write it as the sum over the products of all the elements but including an N-dimensional version of the totally antisymmetric symbol ε_{ijk} introduced in Section 4.1.2.1.

Let's use this to get the idea. If $\underline{\underline{A}}$ a 1×1 matrix then the determinant is just the single element A_{11}. For a 2×2 matrix, it's more complicated but still pretty simple.

[3]If I were writing these notes to get students ready for Quantum Mechanics, I would have used the "dagger" notation instead of the "tilde" notation for dual vectors and Hermitian transpose matrices. That is, here I am writing $\tilde{\underline{\underline{A}}}$ instead of $\underline{\underline{A}}^{\dagger}$. However, these concepts are more generally useful and the only place I'm aware that "daggers" are used is in Quantum Mechanics.

[4]Never confuse $|\underline{\underline{A}}|$ with the concept of "absolute value." Sometimes we write $\det\underline{\underline{A}}$ for $|\underline{\underline{A}}|$.

We just have to multiply along the left and right diagonals, and include the minus sign. That is

$$|\underline{\underline{A}}| = \begin{vmatrix} A_{11} & A_{12} \\ A_{21} & A_{22} \end{vmatrix} = A_{11}A_{22} - A_{12}A_{21}$$

For a 3×3 matrix, it gets a little hairy, but let's take it slowly. Go across the top row, and with each element, form the product with the remaining rows and columns. There are two choices for each element in the top row. Remembering to alternate signs, you get

$$\begin{vmatrix} A_{11} & A_{12} & A_{13} \\ A_{21} & A_{22} & A_{23} \\ A_{31} & A_{32} & A_{33} \end{vmatrix} = \begin{array}{c} A_{11}A_{22}A_{33} - A_{11}A_{23}A_{32} \\ -A_{12}A_{21}A_{33} + A_{12}A_{23}A_{31} \\ +A_{13}A_{21}A_{32} - A_{13}A_{22}A_{31} \end{array}$$

This determinant can be rewritten neatly in terms of 2×2 determinants as

$$\begin{vmatrix} A_{11} & A_{12} & A_{13} \\ A_{21} & A_{22} & A_{23} \\ A_{31} & A_{32} & A_{33} \end{vmatrix} = A_{11}\begin{vmatrix} A_{22} & A_{23} \\ A_{32} & A_{33} \end{vmatrix} - A_{12}\begin{vmatrix} A_{21} & A_{23} \\ A_{31} & A_{33} \end{vmatrix} + A_{13}\begin{vmatrix} A_{21} & A_{22} \\ A_{31} & A_{32} \end{vmatrix}$$

It is easy to see that the three 2×2 matrices in this formula are obtained by removing from the original matrix the row and column corresponding to the top row element in question.

I won't prove it, but I think you can believe that this procedure extends to $N \times N$ matrices. That is, go along the top row (or any other row, for that matter), select the elements one by one, then multiply that element by the determinant of the sub-matrix obtained by removing the row and column of the element. This corresponding $(N-1) \times (N-1)$ determinant, along with the appropriate sign, is called the *cofactor* of the element you selected. This procedure is called a *cofactor expansion* or *Laplace's expansion*.

This is enough to write down some important properties of determinants. I won't prove any of them, but hopefully they will seem at least plausible to you.

1. If you interchange any two rows or columns of a matrix, then the determinant changes sign. An obvious corollary is that the determinant $|\underline{\underline{A}}| = 0$ for any matrix $\underline{\underline{A}}$ that has two identical rows or columns.
2. If any one row or column of a matrix $\underline{\underline{A}}$ is multiplied by a constant c, then the determinant is $c|\underline{\underline{A}}|$. It therefore follows that if any one row (column) of a matrix is proportional to any other row (column), then $|\underline{\underline{A}}| = 0$.
3. If a matrix $\underline{\underline{A}}$ is multiplied by a constant c, then the determinant is multiplied by c^N, that is $|c\underline{\underline{A}}| = c^N|\underline{\underline{A}}|$.
4. The determinant of the transpose of a matrix is the same as the determinant of the original matrix, that is $|\underline{\underline{A}}^{\mathsf{T}}| = |\underline{\underline{A}}|$.
5. The determinant of the Hermitian transpose of a matrix equals the complex conjugate of the determinant of the original matrix, that is $|\tilde{\underline{\underline{A}}}| = |\underline{\underline{A}}|^*$.
6. For two matrices $\underline{\underline{A}}$ and $\underline{\underline{B}}$, $|\underline{\underline{A}}\,\underline{\underline{B}}| = |\underline{\underline{A}}||\underline{\underline{B}}| = |\underline{\underline{B}}\,\underline{\underline{A}}|$. That is, the determinant of the product of matrices is the product of the determinants, regardless of whether or not $\underline{\underline{A}}$ and $\underline{\underline{B}}$ commute.

We will rely on the properties of the determinant much more than actually calculating determinants. In any case, as I mentioned, nobody really calculates determinants anymore, much the same as that nobody calculates square roots anymore. We leave these to computer applications now.

The most important reason for us to know about the determinant is because it is needed to predict properties of matrix inversion. We take that up now.

6.3.6 MATRIX INVERSION

Another important result that we are not going to prove is that the elements of the inverse $\underline{\underline{A}}^{-1}$ of a matrix $\underline{\underline{A}}$ are given by

$$\left(\underline{\underline{A}}^{-1}\right)_{ij} = \frac{1}{|\underline{\underline{A}}|}\left(\underline{\underline{C}}^{\mathrm{T}}\right)_{ij} \tag{6.17}$$

where $\underline{\underline{C}}$ is the matrix of cofactors. That is, an element C_{ij} of $\underline{\underline{C}}$ is just the cofactor you get when you remove the ith row and jth column of $\underline{\underline{A}}$. As with the determinant, we rarely actually calculate the inverse matrix anymore, and leave that up to computer applications.

The important point from (6.17) is that if $|\underline{\underline{A}}| = 0$ then there is no inverse. As we mentioned in the properties of the determinant, this happens if any two rows or columns are identical, or if any row or column is just a factor times another row or column.

The inverse of a diagonal matrix is the diagonal matrix of the inverse of each of the diagonal elements in order. In addition to sounding like this makes sense, it is easy to show. Writing $A_{ij} = a^{(i)}\delta_{ij}$, then we are saying that $A^{-1}_{ij} = [1/a^{(i)}]\delta_{ij}$ so writing it out we get

$$\left(\underline{\underline{A}}^{-1}\underline{\underline{A}}\right)_{ij} = A^{-1}_{ik}A_{kj} = \frac{1}{a^{(i)}}\delta_{ik}a^{(j)}\delta_{kj} = \frac{a^{(j)}}{a^{(i)}}\delta_{ij} = \left(\underline{\underline{I}}\right)_{ij}$$

6.3.7 ORTHOGONAL, HERMITIAN, AND UNITARY MATRICES

This section is just to define three types of matrices. I will give you some indication of why they are important for different physical situations.

A matrix $\underline{\underline{A}}$ is said to be *orthogonal* if its transpose equals its inverse, that is $\underline{\underline{A}}^{\mathrm{T}} = \underline{\underline{A}}^{-1}$. The name comes from the fact that these matrices create rotations in two or three dimensions, and rotations maintain the orthogonality (and norm) of vectors. Note that the elements of such rotation matrices are real numbers, so there is no difference between the transpose and Hermitian transpose for rotation matrices.

Let's see how this works in the case of rotations in two dimensions. Recall the rotation matrix which we called $\underline{\underline{R}}$ in (6.10). For a rotation through an angle ϕ, the relevant matrix is

$$\underline{\underline{A}} = \begin{bmatrix} \cos\phi & -\sin\phi \\ \sin\phi & \cos\phi \end{bmatrix} \tag{6.18}$$

The inverse matrix $\underline{\underline{A}}^{-1}$ is simply obtained by taking $\phi \to -\phi$, so we have

$$\underline{\underline{A}}^{-1} = \begin{bmatrix} \cos\phi & \sin\phi \\ -\sin\phi & \cos\phi \end{bmatrix} = \underline{\underline{A}}^{\mathrm{T}}$$

(It is simple to prove this is the right inverse, just by multiplying out $\underline{\underline{A}}^{-1}\underline{\underline{A}}$.)

A *Hermitian matrix* is one for which the Hermitian transpose leaves it unchanged, that is $\tilde{\underline{\underline{A}}} = \underline{\underline{A}}$. In Quantum Mechanics, Hermitian matrices represent measurable quantities aka "observables." This is closely tied to the fact that Hermitian matrices have real eigenvalues, as we will discuss in Section 6.4.

If a matrix $\underline{\underline{U}}$ has the property that its Hermitian transpose equals its inverse, that is $\tilde{\underline{\underline{U}}} = \underline{\underline{U}}^{-1}$, then we say the matrix is *unitary*. Unitary matrices are practically useful for changing from one basis of a vector space to another. All real, orthogonal matrices are obviously unitary, and are, in fact, useful for changing the axes (i.e. "basis") for locations in two- or three-dimensional space.

6.3.8 CLASSIFYING MATRICES WITH GROUPS

I mentioned back in Section 1.1.4 that physicists make use of groups to quantify symmetry in physical theory. Probably your most practical example of this has to do with matrices.

We've just seen an example. The matrices given by (6.18) form a group, under the binary operation of matrix multiplication. The inverse is given, and the identity element is just the 2×2 identity matrix. The group has an infinite number of elements because ϕ is a continuous variable, but there's nothing wrong with that.

In other words, the matrices (6.18) measure the symmetry corresponding to rotations in a plane. This group has a name, namely SO(2), which stands for the "special orthogonal group in two dimensions." These are "special" matrices because they are constrained to have determinant equal to unity.

Similarly, the group of rotations in three dimensions is measured by the symmetry group SO(3). This group is more complicated to write down because it requires three (real) angles to specify a rotation in three dimensions. A closely connected (but not one-to-one correspondent) group is SU(2), the group of special unitary matrices in two dimensions.

The complex numbers provide an interesting illustration of the concept of *isomorphism* between groups. If we write a complex number as

$$a + ib = re^{i\theta} = r[\cos\theta + i\sin\theta]$$

then it is clear that there is a one-to-one correspondence to the group SO(2). That is

$$a + ib \qquad \Longleftrightarrow \qquad \begin{bmatrix} a & -b \\ b & a \end{bmatrix} \tag{6.19}$$

where $a = r\cos\theta$, $b = r\sin\theta$, and $\theta = -\phi$ makes clear the correspondence with (6.18), and the overall factor of r is irrelevant for our purposes here. In other words, either form in (6.19) can be used to represent a complex number. It is worth pointing out that any 2×2 matrix of this form commutes with any other, that is

$$\begin{bmatrix} a & -b \\ b & a \end{bmatrix} \begin{bmatrix} c & -d \\ d & c \end{bmatrix} = \begin{bmatrix} c & -d \\ d & c \end{bmatrix} \begin{bmatrix} a & -b \\ b & a \end{bmatrix} = \begin{bmatrix} ac-bd & -ad-bc \\ bc+ad & ac-bd \end{bmatrix}$$

and the multiplication of these matrices is equivalent to the multiplication of the complex numbers.

Now, the "1×1" matrices $e^{i\theta}$ are unitary, and in fact form a group called $U(1)$. This all means that there is a one-to-one correspondence between the elements of $U(1)$ and the elements of SO(2). In group theory, one says that SO(2) and U(1) are isomorphic to each other.

6.3.9 REVISITING SYSTEMS OF LINEAR EQUATIONS

We have now gone through all of the necessary material to justify the methodology outlined in Section 6.1 on solving systems of linear algebraic equations. I introduced the subject of Linear Algebra this way so that you could immediately see the practical importance of the formalism. This formalism is, however, useful for a very wide range of physics problems. We will begin to see that now, as we study the formalism of eigenvalues and eigenvectors.

Maybe this is a good time to remind you that we've in fact already used this formalism in Section 3.1.1, where we came up with the Wronskian as defined in (3.5) to see if solutions to a linear ordinary differential equation were independent. The idea was that the ability to satisfy the boundary conditions on the function and its derivative(s) meant that we had to solve a system of linear equations. Linear independence meant that that system had to have a solution, which required the determinant of the coefficients, that is the Wronskian, to be nonzero.

6.4 THE EIGENVALUE PROBLEM

A matrix $\underline{\underline{A}}$ operates on a vector $\underline{v}$ turning into another vector $\underline{\underline{u}}$. Suppose (for reasons that will become clear shortly) we ask ourselves what it means if $\underline{\underline{u}}$ is the same vector as $\underline{\underline{v}}$, perhaps multiplied by some constant λ. That is

$$\underline{\underline{A}}\,\underline{v} = \lambda\,\underline{v} \tag{6.20}$$

In fact, this situation arises very often while solving problems in the physical sciences. We saw it, for example, in Section 3.7. (I'll be more explicit in Section 6.4.3.)

Equation (6.20) is called an *eigenvalue equation*. For a given matrix $\underline{\underline{A}}$, the constant λ is called an *eigenvalue* and the vector $\underline{v}$ is called an *eigenvector*.

Notice that I can multiply (6.20) through by a constant c, in which the eigenvector would be $c\underline{v}$, so there is clearly some freedom in choosing the norm of $\underline{v}$. In most problems, we agree that $\underline{v}$ should have unit norm, and that determines the value of $c = 1/\langle v|v\rangle^{1/2}$.

We will see that an $N \times N$ matrix has N eigenvalues and N eigenvectors, one to go with each of the eigenvalues. It is possible that two or more eigenvalues will be

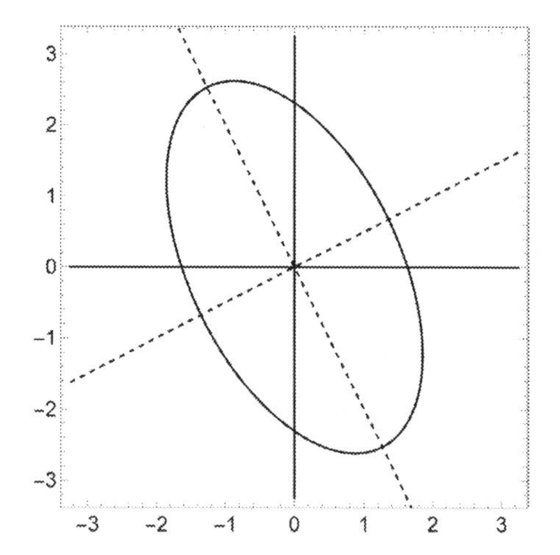

Figure 6.2 The black curve shows the "tilted ellipse" $6x^2 + 4xy + 3y^2 = 16$. The object of this exercise is to find the directions of the major and minor axes of the ellipse, drawn here in dashed lines. In fact, this problem is neatly solved using the eigenvalue approach, resulting in a transformation that "diagonalizes" the matrix used to write the left hand side of the equation. Details are in Section 6.4.1.

equal to each other, and that introduces some complications. These complications are easily overcome, but we won't bother with these situations in this course.

I will illustrate the fundamentals and usefulness of the eigenvalue problem by giving a specific example, namely finding the axes of a tilted ellipse. Along the way we'll prove a general theorem or two that help show why this is such an important problem. After the tilted ellipse, we'll get into the nitty gritty of how to solve the eigenvalue problem in general, that is, finding the eigenvalues and eigenvectors.

6.4.1 THE AXES OF A TILTED ELLIPSE

Figure 6.2 plots the points which satisfy the equation

$$6x^2 + 4xy + 3y^2 = 16 \tag{6.21}$$

which you probably know is the equation of an ellipse. The presence of the "cross term" proportional to xy means that the ellipse is "tilted." That is, its axes do not line up with the x- and y-axes, as ellipses are usually drawn.

Our job is to find the directions of the axes (shown as dashed lines in Figure 6.2) along which the major and minor axes lie. We'll do this by looking for the rotation matrix $\underline{\underline{R}}$ that rotates the x- and y-axes into new axes x' and y' so that the cross terms in (6.21) vanish when written in terms of x' and y' instead of x and y.

Of course, this can be accomplished by writing down rotated x' and y' axes in terms of a rotation angle θ, and then choosing θ so that the cross terms are eliminated. (See the problems at the end of this chapter.) However, we'll take a more sophisticated approach here, to illustrate the eigenvalue technique.

The first step is to write (6.21) in terms of vectors and matrices. This is easy, that is

$$\underline{x}^{\mathsf{T}}\underline{\underline{A}}\,\underline{x} = 16 \qquad \text{where} \qquad \underline{\underline{A}} = \begin{bmatrix} 6 & 2 \\ 2 & 3 \end{bmatrix} \qquad \text{and} \qquad \underline{x} = \begin{bmatrix} x \\ y \end{bmatrix} \tag{6.22}$$

I am using the transpose instead of the Hermitian transpose because these are all real matrices, so that is simpler, and transpose and Hermitian transpose are therefore the same.

I had other choices for $\underline{\underline{A}}$, but I chose this form because it is a symmetric matrix. You'll see soon why that is the choice I had to make.

Now we are looking for the real, orthogonal matrix $\underline{\underline{R}}$ where the transformation

$$\underline{x} = \underline{\underline{R}}\,\underline{x}' \tag{6.23}$$

gets rid of the cross terms in the equation of the ellipse. (Technically, we are looking for the transformation that takes $\underline{x}$ to $\underline{x}'$, but that's just the inverse transformation, given by the transpose of $\underline{\underline{R}}$.) Since the transformation equation implies that $\underline{x}^{\mathrm{T}} = (\underline{x}')^{\mathrm{T}}\underline{\underline{R}}^{\mathrm{T}}$, the ellipse equation (6.22) becomes

$$(\underline{x}')^{\mathrm{T}}\underline{\underline{R}}^{\mathrm{T}}\underline{\underline{A}}\,\underline{\underline{R}}\,\underline{x}' = 16 \tag{6.24}$$

There are no cross terms in (6.24) if the matrix $\underline{\underline{R}}^{\mathrm{T}}\underline{\underline{A}}\,\underline{\underline{R}}$ is diagonal. Therefore, our job reduces, mathematically, to finding the orthogonal matrix $\underline{\underline{R}}$ that "diagonalizes" the symmetric matrix $\underline{\underline{A}}$.

The eigenvalue problem solves this problem for us, because of a very important theorem that I will now state and prove. **The eigenvectors of a Hermitian matrix form an orthogonal set.** A byproduct of this proof will be a proof that the eigenvalues of a Hermitian matrix are real numbers.

The proof starts here. Consider two eigenvalues $\lambda^{(a)}$ and $\lambda^{(b)}$ and their corresponding eigenvectors $\underline{v}^{(a)}$ and $\underline{v}^{(b)}$. The eigenvalue equations are

$$\underline{\underline{A}}\,\underline{v}^{(a)} = \lambda^{(a)}\underline{v}^{(a)} \qquad \text{and} \qquad \underline{\underline{A}}\,\underline{v}^{(b)} = \lambda^{(b)}\underline{v}^{(b)}$$

Now take the inner product from the left with $\underline{v}^{(b)}$ on the first equation, and with $\underline{v}^{(a)}$ on the second equation. This gives

$$\tilde{\underline{v}}^{(b)}\underline{\underline{A}}\,\underline{v}^{(a)} = \lambda^{(a)}\tilde{\underline{v}}^{(b)}\,\underline{v}^{(a)} \qquad \text{and} \qquad \tilde{\underline{v}}^{(a)}\underline{\underline{A}}\,\underline{v}^{(b)} = \lambda^{(b)}\tilde{\underline{v}}^{(a)}\,\underline{v}^{(b)} \tag{6.25}$$

Now look at the left hand side of the first equation. This is a number that we can write as

$$\tilde{\underline{v}}^{(b)}\underline{\underline{A}}\,\underline{v}^{(a)} = v_i^{(b)^*}A_{ij}v_j^{(a)} = \left(v_i^{(b)}A_{ij}^{*}v_j^{(a)^*}\right)^* = \left(v_j^{(a)^*}\tilde{A}_{ji}v_i^{(b)}\right)^* = \left(\tilde{\underline{v}}^{(a)}\tilde{\underline{\underline{A}}}\,\underline{v}^{(b)}\right)^*$$

For the right side of the first equation, we have something similar, namely

$$\tilde{\underline{v}}^{(b)}\,\underline{v}^{(a)} = v_i^{(b)^*}v_i^{(a)} = \left(v_i^{(b)}v_i^{(a)^*}\right)^* = \left(\tilde{\underline{v}}^{(a)}\,\underline{v}^{(b)}\right)^*$$

In other words, flipping the order of these inner products means to take the complex conjugate of the result, and replacing the matrix by its Hermitian conjugate. This all means that we can rewrite (6.25) by using the above relationships and taking the complex conjugate of the first equation and get

$$\tilde{\underline{v}}^{(a)}\tilde{\underline{\underline{A}}}\,\underline{v}^{(b)} = \lambda^{(a)^*}\tilde{\underline{v}}^{(a)}\,\underline{v}^{(b)} \qquad \text{and} \qquad \tilde{\underline{v}}^{(a)}\underline{\underline{A}}\,\underline{v}^{(b)} = \lambda^{(b)}\tilde{\underline{v}}^{(a)}\,\underline{v}^{(b)} \tag{6.26}$$

If $\underline{\underline{A}}$ is Hermitian, that is $\tilde{\underline{\underline{A}}} = \underline{\underline{A}}$, then the left sides of these two equations are the same. Subtracting them tells us that

$$0 = \left(\lambda^{(a)^*} - \lambda^{(b)}\right) \tilde{\underline{v}}^{(a)} \underline{v}^{(b)}$$

Now if $a = b$, then the inner product $\tilde{\underline{v}}^{(a)} \underline{v}^{(b)} = \tilde{\underline{v}}^{(a)} \underline{v}^{(a)}$ is positive definite, so

$$\lambda^{(a)^*} = \lambda^{(a)}$$

proving that the eigenvalues are real. On the other hand, if $a \neq b$, and the eigenvalues are distinct, then

$$\tilde{\underline{v}}^{(a)} \underline{v}^{(b)} = 0$$

and the eigenvectors are orthogonal. (As I mentioned earlier, I am leaving the case of different eigenvectors having the same eigenvalue to a more advanced course.) **This completes the proof.**

One immediate consequence of this proof is that the eigenvectors of a Hermitian matrix form a *complete set*. That is, any arbitrary vector $\underline{x}$ can be written as a linear combination of the eigenvectors. This is easy to see, if we just assume that

$$\underline{x} = \sum_{i=1}^{N} c_i \underline{v}^{(i)}$$

and show that we can come up with a formula for the c_i. However, it is clear that

$$\langle v^{(j)} | x \rangle = \sum_{i=1}^{N} c_i \langle v^{(j)} | v^{(i)} \rangle = \sum_{i=1}^{N} c_i \delta_{ij} = c_j$$

Therefore, the desired formula for the c_i is simply

$$c_i = \langle v^{(i)} | x \rangle \qquad \text{that is} \qquad \underline{x} = \sum_{i=1}^{N} \langle v^{(i)} | x \rangle \underline{v}^{(i)}$$

Now let's see how the properties of Hermitian matrices help us figure out the axes of a tilted ellipse. Our matrix $\underline{\underline{A}}$ is real and symmetric, so it is also Hermitian. Therefore, the two eigenvectors of $\underline{\underline{A}}$ will be orthogonal to each other.

This tells us how to build $\underline{\underline{R}}$, namely by constructing the matrix whose columns are the (normalized) eigenvectors of $\underline{\underline{A}}$, that is

$$\underline{\underline{R}} = \left[\underline{v}^{(1)} \;\; \underline{v}^{(2)} \;\; \cdots \;\; \underline{v}^{(N)}\right] \tag{6.27}$$

Firstly, this ensures that $\underline{\underline{R}}$ is an orthogonal matrix, $\underline{\underline{R}}^\mathsf{T}\underline{\underline{R}} = \underline{\underline{I}}$. Now, when we do $\underline{\underline{A}}\,\underline{\underline{R}}$, then, each column will be just multiplied by the eigenvalue. So when we do $\underline{\underline{R}}^\mathsf{T}$ on this result, that is we calculate $\underline{\underline{R}}^\mathsf{T}\underline{\underline{AR}}$, we will get a diagonal matrix with the eigenvalues along the diagonal! This is a general argument that will work any $N \times N$ matrix $\underline{\underline{A}}$.

This is all best illustrated by going back to our tilted ellipse problem. After doing the work to get the eigenvalues and eigenvectors of $\underline{\underline{A}}$ in (6.22), a procedure we will describe in detail in Section 6.4.2, we end up with eigenvalues

$$\lambda^{(1)} = 7 \qquad \text{and} \qquad \lambda^{(2)} = 2$$

corresponding to the (normalized) eigenvectors

$$\underline{v}^{(1)} = \begin{bmatrix} 2/\sqrt{5} \\ 1/\sqrt{5} \end{bmatrix} \qquad \text{and} \qquad \underline{v}^{(2)} = \begin{bmatrix} -1/\sqrt{5} \\ 2/\sqrt{5} \end{bmatrix}$$

Using the eigenvectors as columns in the rotation matrix gives

$$\underline{\underline{R}} = \begin{bmatrix} 2/\sqrt{5} & -1/\sqrt{5} \\ 1/\sqrt{5} & 2/\sqrt{5} \end{bmatrix} \qquad \text{and, so} \qquad \underline{\underline{R}}^{\mathsf{T}} = \begin{bmatrix} 2/\sqrt{5} & 1/\sqrt{5} \\ -1/\sqrt{5} & 2/\sqrt{5} \end{bmatrix}$$

It is worth showing explicitly that $\underline{\underline{R}}$ is orthogonal, so let's do it.

$$\begin{aligned} \underline{\underline{R}}^{\mathsf{T}}\underline{\underline{R}} &= \begin{bmatrix} 2/\sqrt{5} & 1/\sqrt{5} \\ -1/\sqrt{5} & 2/\sqrt{5} \end{bmatrix} \begin{bmatrix} 2/\sqrt{5} & -1/\sqrt{5} \\ 1/\sqrt{5} & 2/\sqrt{5} \end{bmatrix} \\ &= \begin{bmatrix} 4/5+1/5 & -2/5+2/5 \\ -2/5+1/5 & 1/5+4/5 \end{bmatrix} = \begin{bmatrix} 1 & 0 \\ 0 & 1 \end{bmatrix} = \underline{\underline{I}} \end{aligned}$$

You should take a moment to convince yourself that $\underline{\underline{R}}\underline{\underline{R}}^{\mathsf{T}} = \underline{\underline{I}}$ as well.

Now let's check our conjecture that building $\underline{\underline{R}}$ this way "diagonalizes" the matrix $\underline{\underline{A}}$. We'll do this in two steps for the sake of illustration. First we multiply $\underline{\underline{A}}$ times $\underline{\underline{R}}$, so

$$\underline{\underline{A}}\underline{\underline{R}} = \begin{bmatrix} 6 & 2 \\ 2 & 3 \end{bmatrix} \begin{bmatrix} 2/\sqrt{5} & -1/\sqrt{5} \\ 1/\sqrt{5} & 2/\sqrt{5} \end{bmatrix} = \begin{bmatrix} 14/\sqrt{5} & -2/\sqrt{5} \\ 7/\sqrt{5} & 4/\sqrt{5} \end{bmatrix}$$

As predicted, this just multiplies each column by its respective eigenvalue. Of course, it had to be, because we built $\underline{\underline{R}}$ with columns that in fact were the eigenvectors of $\underline{\underline{A}}$.

Second, we complete the transformation indicated in (6.24) to get

$$\begin{aligned} \underline{\underline{R}}^{\mathsf{T}}\underline{\underline{A}}\underline{\underline{R}} &= \begin{bmatrix} 2/\sqrt{5} & 1/\sqrt{5} \\ -1/\sqrt{5} & 2/\sqrt{5} \end{bmatrix} \begin{bmatrix} 14/\sqrt{5} & -2/\sqrt{5} \\ 7/\sqrt{5} & 4/\sqrt{5} \end{bmatrix} \\ &= \begin{bmatrix} (28+7)/5 & (-4+4)/5 \\ (-14+14)/5 & (2+8)/5 \end{bmatrix} = \begin{bmatrix} 7 & 0 \\ 0 & 2 \end{bmatrix} \end{aligned}$$

and, again as predicted, the transformed matrix is diagonal with elements given by the eigenvalues. Therefore, writing out (6.24) in terms of our tilted x'- and y'-axes,

$$7x'^2 + 2y'^2 = 16$$

which is the form we were aiming to achieve.

If we want to find the actual rotation angle that we used, just interpret the element R_{11} as the cosine of the angle. That is

$$\phi = \cos^{-1}\frac{2}{\sqrt{5}} = 26.6^\circ$$

The dashed lines in Figure 6.2 are given by the eigenvectors themselves. That is, the directions of these lines are given by the eigenvectors. To see this, first recall from Section 4.1.4 that the formula for a straight line passing through the origin, in terms of a real parameter t, is $\vec{r} = \vec{m}t$ where $\vec{m}$ gives the direction of the line. Now, the directions of the symmetry axes of the ellipse are given by unit vectors in the (x', y', z') system. If we wrote this as an N-component generalized column vector, then it would have a "one" in some row, and "zero" in all of the others. If we want to express this direction in the (x, y, z) system, then use (6.23) with (6.27). This should make it clear that each axis direction vector in the (x', y', z') becomes an eigenvector in the (x, y, z) system.

We can illustrate this with our two-dimensional rotated ellipse. The two eigenvectors, written as two-dimension vectors in the (x, y) system, ignoring the $\sqrt{5}$ in the denominator, are

$$\vec{m}^{(1)} = 2\hat{\imath} + \hat{k} \qquad \text{and} \qquad \vec{m}^{(2)} = -\hat{\imath} + 2\hat{k}$$

The first of these corresponds to the line

$$x = 2t \qquad \text{and} \qquad y = t \qquad \text{or} \qquad x = 2y$$

The second corresponds to the line

$$x = -t \qquad \text{and} \qquad y = 2t \qquad \text{or} \qquad y = -2x$$

These are the two dashed lines in Figure 6.2.

To summarize, after proving a theorem about the eigenvectors of Hermitian matrices, we used the results of that theorem to build an orthogonal matrix $\underline{\underline{R}}$ that "diagonalized" the matrix $\underline{\underline{A}}$ which solved our problem of finding the tilted axes of an ellipse. It should be clear, however, that this approach would work for *any* $N \times N$ Hermitian (or real symmetric) matrix $\underline{\underline{A}}$. This touches on very many important physical problems.

Now that we see how this works, let's discuss how to actually go about finding the eigenvectors and eigenvalues of a matrix.

6.4.2 FINDING EIGENVALUES AND EIGENVECTORS

First, I will own up to the fact that to find the eigenvalues and eigenvectors of the matrix $\underline{\underline{A}}$ in (6.22), I used a computer application. Just as "nobody" calculates determinants anymore, "nobody" does the eigenvector calculation by hand. Nevertheless, we'll go through the procedure, just to show you that it is not magic. We'll also illustrate it with our 2×2 matrix example.

We first rewrite (6.20) slightly as

$$\underline{\underline{A}}\,\underline{v} = \lambda\,\underline{\underline{I}}\,\underline{v}$$

where we have inserted the identity matrix in front of $\underline{v}$, that is we've used $\underline{\underline{I}}\,\underline{v} = \underline{v}$. With a little bit of rearranging, this gives

$$\left(\underline{\underline{A}} - \lambda\,\underline{\underline{I}}\right)\underline{v} = \underline{0} \tag{6.28}$$

where we are being explicit that the right side of the equation is the column vector with all entries equal to zero. This is just a system of linear algebraic equations for the components of $\underline{v}$. What's more, it is a *homogeneous* system of equations. That means that we expect all of the components of $\underline{v}$ to be zero.

This is unacceptable, of course, so we need to prevent (6.28) from having a solution. We know how to do this, though. We just require that the determinant of the matrix on the left be zero. Mathematically, this means

$$\det\left(\underline{\underline{A}} - \lambda\,\underline{\underline{I}}\right) = \begin{vmatrix} A_{11}-\lambda & A_{12} & A_{13} & \cdots & A_{1N} \\ A_{21} & A_{22}-\lambda & A_{23} & \cdots & A_{2N} \\ A_{31} & A_{32} & A_{33}-\lambda & \cdots & A_{3N} \\ \vdots & \vdots & \vdots & & \vdots \\ A_{N1} & A_{N2} & A_{N3} & \cdots & A_{NN}-\lambda \end{vmatrix} = 0 \qquad (6.29)$$

which is a polynomial of degree N in λ. This is called the *characteristic equation* and will have N roots. That is, it yields N values for λ.

Each value of λ makes (6.28) a different set of equations that can be solved for the components of $\underline{v}$. Though this system is no longer N independent equations, so the best you can do is to solve for $N-1$ components in terms of the one remaining. That's OK, though, because you want to normalize $\underline{v}$, giving you an additional equation, namely $\langle v|v\rangle = 1$.

It's best to illustrate this with a specific example, so let's use the matrix $\underline{\underline{A}}$ from (6.22). The characteristic equation is

$$\begin{vmatrix} 6-\lambda & 2 \\ 2 & 3-\lambda \end{vmatrix} = (6-\lambda)(3-\lambda) - 4 = \lambda^2 - 9\lambda + 14 = (\lambda-7)(\lambda-2) = 0$$

so the eigenvalues are indeed 7 and 2. For $\lambda = 7$, the system of equations is

$$\begin{aligned} -v_1 + 2v_2 &= 0 \\ 2v_1 - 4v_2 &= 0 \end{aligned}$$

which are indeed the same equation, which reduces to $v_1 = 2v_2$. Combining this with

$$\langle v|v\rangle = v_1^2 + v_2^2 = 5v_2^2 = 1$$

gives us $v_1 = 2/\sqrt{5}$ and $v_2 = 1/\sqrt{5}$ which is what we quoted for the eigenvector $\underline{v}$ corresponding to the eigenvalue $\lambda = 7$. For $\lambda = 2$, the system of equations is

$$\begin{aligned} 4v_1 + 2v_2 &= 0 \\ 2v_1 + v_2 &= 0 \end{aligned}$$

so $v_2 = -2v_1$ and so forth.

For matrices larger than 2×2, the characteristic equation can be tricky to solve, but the procedure is the same.

6.4.3 COUPLED OSCILLATIONS REVISITED

We studied the problem of two masses and three springs in Section 3.7.1. In fact, that is an ideal example of an eigenvalue problem in physics, although we didn't call it that at a time. Formulating that problem in terms of eigenvectors is elegant and straightforward to generalize to N masses.

In this section, we will reformulate the problem of two identical masses connected by three identical springs in terms of vectors, matrices, and eigenvalues. First let's rewrite the coupled differential equations (3.40) as

$$\ddot{x}_1(t) = -2\omega_0^2 x_1(t) + \omega_0^2 x_2(t) \tag{6.30a}$$

$$\ddot{x}_2(t) = \omega_0^2 x_1(t) - 2\omega_0^2 x_2(t) \tag{6.30b}$$

In terms of vectors and matrices, we can write this as

$$\ddot{\underline{x}}(t) = -\omega_0^2 \underline{\underline{\Omega}}\,\underline{x}(t) \tag{6.31}$$

where we have made the definitions

$$\underline{x} = \begin{bmatrix} x_1 \\ x_2 \end{bmatrix} \quad \text{and} \quad \underline{\underline{\Omega}} = \begin{bmatrix} 2 & -1 \\ -1 & 2 \end{bmatrix}$$

It should be clear that if the masses and springs were not all the same, then the formulation would look the same, but the matrix $\underline{\underline{\Omega}}$ would be different.

Now we make our standard ansatz, which now takes the form

$$\underline{x}(t) = \underline{a}\,e^{i\omega t} \quad \text{where} \quad \underline{a} = \begin{bmatrix} a_1 \\ a_2 \end{bmatrix} \tag{6.32}$$

is a vector of constants a_1 and a_2. Taking the time derivative and dividing out the factor $e^{i\omega t}$ on both sides, the vector differential equation (6.31) now takes the form

$$\underline{\underline{\Omega}}\,\underline{a} = \lambda \underline{a} \quad \text{where} \quad \lambda = \frac{\omega^2}{\omega_0^2} \tag{6.33}$$

and we have arrived at an eigenvalue problem.

The eigenvalues of $\underline{\underline{\Omega}}$ are easy to determine. The characteristic equation is

$$\begin{vmatrix} 2-\lambda & -1 \\ -1 & 2-\lambda \end{vmatrix} = (2-\lambda)^2 - 1 = 0$$

is easily solved since $2-\lambda = \pm 1$ so $\lambda = 1$ or $\lambda = 3$. As expected an $N \times N = 2 \times 2$ matrix has $N = 2$ eigenvalues. The eigenvectors are simple to find. Let's use a labeling scheme where the first eigenvalue is $\lambda = 1$, and the second eigenvalue is $\lambda = 3$. Then for $\lambda = 1$,

$$\begin{bmatrix} 1 & -1 \\ -1 & 1 \end{bmatrix} \begin{bmatrix} a_1^{(1)} \\ a_2^{(1)} \end{bmatrix} = \begin{bmatrix} 0 \\ 0 \end{bmatrix}$$

which gives two equivalent equations for $a_1^{(1)}$ and $a_2^{(1)}$, namely $a_1^{(1)} - a_2^{(1)} = 0$ or $a_1^{(1)} = a_2^{(1)}$. Therefore, including a normalization, the first eigenvector is

$$\underline{a}^{(1)} = \begin{bmatrix} 1/\sqrt{2} \\ 1/\sqrt{2} \end{bmatrix} \tag{6.34}$$

For the second eigenvalue $\lambda = 3$, we find the eigenvector using

$$\begin{bmatrix} -1 & -1 \\ -1 & -1 \end{bmatrix} \begin{bmatrix} a_1^{(2)} \\ a_2^{(2)} \end{bmatrix} = \begin{bmatrix} 0 \\ 0 \end{bmatrix}$$

which implies that $a_1^{(2)} + a_2^{(2)} = 0$, so the normalized eigenvector is

$$\underline{a}^{(2)} = \begin{bmatrix} 1/\sqrt{2} \\ -1/\sqrt{2} \end{bmatrix} \tag{6.35}$$

Of course, our real goal here is to find the motions of the two masses, that is $x_1(t)$ and $x_2(t)$, or, equivalently (and more succinctly) the vector $\underline{x}(t)$. Writing the solution as the ansatz (6.32) was just a means to a solution. What we have learned is that there are two solutions, one with "eigenfrequency" $\omega = \pm\omega_0 \equiv \pm\omega^{(1)}$ and the other with eigenfrequency $\omega = \pm\omega_0\sqrt{3} \equiv \omega^{(2)}$. The "$\pm$" is an artefact of our solving a second-order differential equation, giving us in fact two solutions for each eigenvalue.

So, the general solution for the motion of the two masses is

$$\underline{x}(t) = c_+^{(1)}\underline{a}^{(1)}e^{i\omega^{(1)}t} + c_-^{(1)}\underline{a}^{(1)}e^{-i\omega^{(1)}t} + c_+^{(2)}\underline{a}^{(2)}e^{i\omega^{(2)}t} + c_-^{(2)}\underline{a}^{(2)}e^{-i\omega^{(2)}t} \tag{6.36}$$

where the four constants $c_+^{(1)}$, $c_-^{(1)}$, $c_+^{(2)}$, and $c_-^{(2)}$ are determined from the four initial conditions, namely the initial positions and velocities of each of the two masses. Look back now at (3.42), and refer to the eigenvectors $\underline{a}^{(1)}$ (6.34) and $\underline{a}^{(2)}$ (6.35). Equation (6.36) is the same as (3.42), where

$$\frac{1}{\sqrt{2}}c_+^{(1)} = a \qquad \frac{1}{\sqrt{2}}c_-^{(1)} = b \qquad \frac{1}{\sqrt{2}}c_+^{(2)} = c \qquad \frac{1}{\sqrt{2}}c_-^{(2)} = d$$

Let's continue with our vector and matrix notation and put in the initial conditions. The initial position and velocity vectors are

$$\underline{x}_0 = \begin{bmatrix} x_{1_0} \\ x_{2_0} \end{bmatrix} \quad \text{and} \quad \underline{v}_0 = \begin{bmatrix} v_{1_0} \\ v_{2_0} \end{bmatrix}$$

so applying (6.36) gives us

$$\begin{aligned} \underline{x}_0 &= c_+^{(1)}\underline{a}^{(1)} + c_-^{(1)}\underline{a}^{(1)} + c_+^{(2)}\underline{a}^{(2)} + c_-^{(2)}\underline{a}^{(1)} \\ &= \left[c_+^{(1)} + c_-^{(1)}\right]\underline{a}^{(1)} + \left[c_+^{(2)} + c_-^{(2)}\right]\underline{a}^{(2)} \\ \text{and} \quad \underline{v}_0 &= i\omega^{(1)}c_+^{(1)}\underline{a}^{(1)} - i\omega^{(1)}c_-^{(1)}\underline{a}^{(1)} + i\omega^{(2)}c_+^{(2)}\underline{a}^{(2)} - i\omega^{(2)}c_-^{(2)}\underline{a}^{(1)} \\ &= i\omega^{(1)}\left[c_+^{(1)} - c_-^{(1)}\right]\underline{a}^{(1)} + i\omega^{(2)}\left[c_+^{(2)} - c_-^{(2)}\right]\underline{a}^{(2)} \end{aligned}$$

These equations look messy, but don't let that slow you down. Remember that the $\omega^{(1)}$ and $\omega^{(2)}$ are just numbers, as are the components of $\underline{x}_0$ and $\underline{v}_0$, so this is just four equations to solve for the $c_\pm^{(1)}$ and $c_\pm^{(2)}$ in terms of the other stuff.

If we look at a simple special case, we can get a better feeling for how the motion breaks down in terms of the eigenvectors. Let's say the two masses start from rest, that is $\underline{v}_0 = \underline{0}$. This means that $c_+^{(1)} = c_-^{(1)} \equiv c^{(1)}/2$ and $c_+^{(2)} = c_-^{(2)} \equiv c^{(2)}/2$, and

$$\underline{x}_0 = c^{(1)}\underline{a}^{(1)} + c^{(2)}\underline{a}^{(2)} \tag{6.37}$$

It is now clear how we excite the "eigenmodes." If we set $x_{1_0} = x_{2_0} = c$, that is

$$\underline{x}_0^{(1)} = \begin{bmatrix} c \\ c \end{bmatrix} = c\sqrt{2}\underline{a}^{(1)}$$

which just means that the initial positions of the two masses correspond to the eigenvalue $\lambda^{(1)} = 1$, i.e. $\omega = \omega_0$, then $c^{(2)} = 0$ in order to satisfy (6.37) and the motion (6.36) becomes

$$\underline{x}^{(1)}(t) = \frac{1}{2}c^{(1)}\underline{a}^{(1)}e^{i\omega^{(1)}t} + \frac{1}{2}c^{(1)}\underline{a}^{(1)}e^{-i\omega^{(1)}t} = c^{(1)}\underline{a}^{(1)}\cos\omega^{(1)}t = c\sqrt{2}\underline{a}^{(1)}\cos\omega_0 t$$

If we write this as two separate equations using the eigenvector $\underline{a}^{(1)}$ from (6.34), we have

$$x_1^{(1)}(t) = c\cos\omega_0 t \qquad \text{and} \qquad x_2^{(1)}(t) = c\cos\omega_0 t$$

In other words, in "eigenmode (1)," the two masses oscillate together, in phase, exactly as shown on the left in Figure 3.14.

To excite "eigenmode (2)," we set initial conditions that correspond to eigenvector $\underline{a}^{(2)}$ from (6.35). By setting $x_{1_0} = c$ and $x_{2_0} = c$ we have

$$\underline{x}_0^{(2)} = \begin{bmatrix} c \\ -c \end{bmatrix} = c\sqrt{2}\underline{a}^{(2)}$$

corresponding to $\lambda^{(2)} = 3$, i.e. $\omega = \omega_0\sqrt{3}$. We therefore set $c^{(1)} = 0$ in order to satisfy (6.37) and the motion (6.36) becomes

$$\underline{x}^{(2)}(t) = \frac{1}{2}c^{(2)}\underline{a}^{(2)}e^{i\omega^{(2)}t} + \frac{1}{2}c^{(2)}\underline{a}^{(2)}e^{-i\omega^{(2)}t} = c^{(2)}\underline{a}^{(2)}\cos\omega^{(2)}t = c\sqrt{2}\underline{a}^{(2)}\cos\omega_0\sqrt{3}t$$

Written as two separate equations using the eigenvector $\underline{a}^{(2)}$ from (6.35), we have

$$x_1^{(2)}(t) = c\cos\omega_0\sqrt{3}t \qquad \text{and} \qquad x_2^{(1)}(t) = -c\cos\omega_0\sqrt{3}t$$

In other words, in "eigenmode (2)," the two masses oscillate against each other, 180° out of phase, with a frequency that is $\sqrt{3}$ higher than for eigenmode (1), exactly as shown on the right in Figure 3.14.

To summarize, we have solved the problem of two identical masses connected the three identical springs using the eigenvalue problem formalism, and the result is (of course!) exactly the same as what we got with our seat-of-the-pants approach in Section 3.7.1.

The eigenvalue problem approach gets a little tedious when we actually get down to writing the motions of the two masses, but there is a very important reason why this approach is the better way to go. The differential equation written as (6.31) is much more elegant than writing the two separate equations (3.40), and it is much more easily generalized to systems of different masses or larger numbers of masses. In fact, in most physical problems, it isn't the actual motion of the individual masses that matter, but what are their eigenmodes and eigenfrequencies. This comes directly from constructing the $N \times N$ matrix $\underline{\underline{\Omega}}$ for a given problem with N masses.

6.5 FOUR-VECTORS IN SPACETIME

The hyperbolic functions from Section 1.5.4 have an interesting role to play in the formalism of special relativity. We can form generalized "four vectors" that specify the position of a particle in "spacetime," that is, its position in three-dimensional space as well as its time. The inner product has to be defined in a way that looks peculiar, but that's because we haven't studied the concept of a *metric*.

The point is that "rotations" in this four dimensional space, called Lorentz Transformations, are generated by 4×4 matrices that include hyperbolic sine and cosine instead of circular sine and cosine. I encourage you to look into this more.

EXERCISES

6.1 *Find the norm of the vector* $\underline{u} = \underline{\underline{A}}\,\underline{v}$ *where*

$$\underline{\underline{A}} = \begin{bmatrix} 1 & 0 & i \\ -2 & i & 2 \\ 2 & -2 & 4 \end{bmatrix} \quad \text{and} \quad \underline{v} = \begin{bmatrix} 1 \\ 2 \\ 2 \end{bmatrix}$$

6.2 *Find the inverse* $\underline{\underline{A}}^{-1}$ *for the matrix*

$$\underline{\underline{A}} = \begin{bmatrix} 1 & 1 & 1 \\ 1 & 0 & 1 \\ 1 & 0 & -1 \end{bmatrix}$$

by solving the system of equations $\underline{\underline{A}}\,\underline{x} = \underline{c}$ *for the vector* $\underline{x}$ *in terms of an arbitrary vector* $\underline{c}$ *and expressing your result as* $\underline{x} = \underline{\underline{A}}^{-1}\underline{c}$.

6.3 *Write out the three equations for x, y, and z represented by*

$$\underline{\underline{A}}\underline{X} = \underline{C} \quad \text{where} \quad \underline{\underline{A}} = \begin{bmatrix} 1 & 2 & 4 \\ 2 & 0 & 1 \\ 1 & 1 & 1 \end{bmatrix} \quad \underline{X} = \begin{bmatrix} x \\ y \\ z \end{bmatrix} \quad \underline{C} = \begin{bmatrix} c_1 \\ c_2 \\ c_3 \end{bmatrix}$$

and solve them for x, y, and z. Then determine the matrix $\underline{\underline{A}}^{-1}$ *by writing your answer as* $\underline{X} = \underline{\underline{A}}^{-1}\underline{C}$.

6.4 *Find the inverse matrix* $\underline{\underline{A}}^{-1}$ *for the matrix* $\underline{\underline{A}}$ *in the previous problem by calculating the determinant using an expansion in minors, and then forming the matrix of cofactors.*

6.5 *Construct a* 3×3 *matrix that rotates a three-dimensional vector* $\underline{v}$ *through an angle* θ *about the x-axis, combined with a reflection about the yz-plane, that is, takes* x *to* $-x$*. Pick two specific examples for* $\underline{v}$ *and show that your matrix does what it is supposed to do.*

6.6 *Prove that if* $\underline{\underline{A}}\,\underline{\underline{B}} = \underline{\underline{0}}$ *for two matrices* $\underline{\underline{A}}$ *and* $\underline{\underline{B}}$*, then the determinant of at least one of them must be zero. Find an example, however, of two* 3×3 *matrices that are each nonzero, and in fact do not have any full rows or columns with all zeros, but whose product* $\underline{\underline{A}}\,\underline{\underline{B}} = \underline{\underline{0}}$*.*

6.7 *The trace* $\mathrm{tr}(\underline{\underline{A}})$ *of a matrix* $\underline{\underline{A}}$ *is defined as the sum over the diagonal elements of* $\underline{\underline{A}}$*, that is* $\mathrm{tr}(\underline{\underline{A}}) = A_{ii}$*. Prove that* $\mathrm{tr}(\underline{\underline{A}}\,\underline{\underline{B}}) = \mathrm{tr}(\underline{\underline{B}}\,\underline{\underline{A}})$ *for any two matrices* $\underline{\underline{A}}$ *and* $\underline{\underline{B}}$*, regardless of whether or not they commute.*

6.8 *Consider rotations in 3D space, recalling how we describe rotations in a plane.*

(a) *Find the* 3×3 *matrix* $\underline{\underline{A}}$ *that rotates a vector by* 90° *around the z-axis.*
(b) *Find the* 3×3 *matrix* $\underline{\underline{B}}$ *that rotates a vector by* 90° *around the x-axis.*
(c) *Show that* $\underline{\underline{A}}\,\underline{\underline{B}} \neq \underline{\underline{B}}\,\underline{\underline{A}}$ *by explicit matrix multiplication.*
(d) *For the vector* $\underline{v} = \hat{j}$*, the unit vector in the y-direction, calculate* $\underline{\underline{A}}\,\underline{\underline{B}}\,\underline{v}$ *and* $\underline{\underline{B}}\,\underline{\underline{A}}\,\underline{v}$*. Sketch diagrams that demonstrate these unequal results. (Don't be too concerned about the sign of the rotation angle.)*

6.9 *Find the norms of the vectors* $\underline{v}$ *and* $\underline{u}$ *below, and also show that they are orthogonal to each other. Then find some vector* $\underline{w}$ *with unit norm that is orthogonal to both* $\underline{v}$ *and* $\underline{u}$*.*

$$\underline{v} = \begin{bmatrix} i \\ 1 \\ -1 \end{bmatrix} \qquad \underline{u} = \begin{bmatrix} 2i \\ -2 \\ 0 \end{bmatrix}$$

6.10 *Solve the problem of the rotated ellipse by writing* x *and* y *in terms of* x'*,* y'*, and a rotation angle* ϕ*, and then see what result you get if you eliminate the cross terms.*

6.11 *Use the properties of determinants to prove that* $\underline{\underline{A}}\,\underline{\underline{B}} = \underline{\underline{0}}$ *implies that either* $|\underline{\underline{A}}| = 0$ *or* $|\underline{\underline{B}}| = 0$*, or both, where* $\underline{\underline{0}}$ *is the matrix of all zeros. Demonstrate this with the matrices*

$$\underline{\underline{A}} = \begin{bmatrix} 1 & 1 \\ 2 & 2 \end{bmatrix} \quad \text{and} \quad \underline{\underline{B}} = \begin{bmatrix} a & b \\ -a & -b \end{bmatrix}$$

where a *and* b *can be any complex numbers. Which matrix has zero determinant?*

6.12 *Use the eigenvalue approach to find the symmetry axes of the conic section* $6x^2 + 12xy + y^2 = 16$*, and find the angle they make with the* x, y *axes. What kind of curve is this? A plot would be helpful. Check your answer by rotating to a new set of axes and eliminating the cross term.*

6.13 *A set of points* (x,y) *in a plane is given by* $5x^2 - 4xy + 2y^2 = 30$ *where* x *and* y *are orthogonal coordinates. Use the eigenvalue approach to write this equation as* $ax'^2 + by'^2 = 30$*, with numerical values for* a *and* b*, where the* x'*- and* y'*-axes are rotated with respect to the* x*- and* y*-axes. Find the angle of rotation. Check your answer by rotating to a new set of axes and eliminating the cross term.*

6.14 *Find the eigenvalues and eigenvectors for the matrix*

$$\underline{\underline{\sigma}}_y = \begin{bmatrix} 0 & -i \\ i & 0 \end{bmatrix}$$

one of the three Pauli matrices. *Normalize the eigenvectors and show that they are orthogonal.*

6.15 *Find the eigenvalues, two of which equal each other, of the real symmetric matrix*

$$\underline{\underline{A}} = \begin{bmatrix} 13 & 4 & -2 \\ 4 & 13 & -2 \\ -2 & -2 & 10 \end{bmatrix}$$

You will find you have more freedom than you would have thought. Do you see how to use this freedom to make all three eigenvectors mutually orthogonal?

6.16 *Find the eigenfrequencies and eigenmodes for the mechanical system*

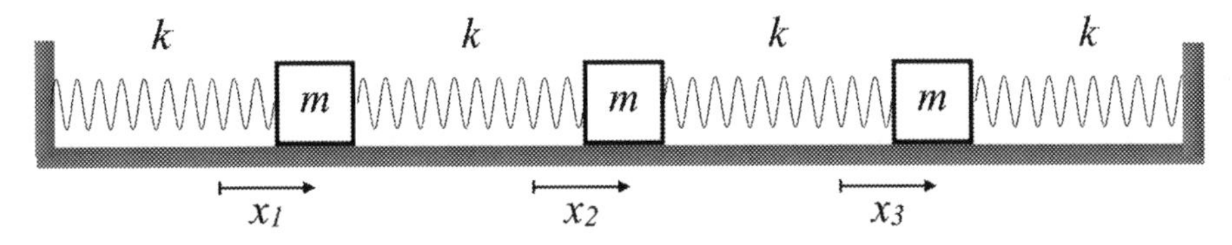

Make a plot that shows the motions of each of the three masses, for the three sets of initial conditions where the masses start at rest with position given by each of the three eigenvectors. Briefly describe the motions of the three masses, for each of the eigenmodes.

6.17 *A matrix* $\underline{\underline{A}}$ *is unitary if* $\tilde{\underline{\underline{A}}} = \underline{\underline{A}}^{-1}$*. Prove that the eigenvalues* λ *of a unitary matrix must be of the form* $\lambda = e^{i\phi}$ *where* ϕ *is a real number. (We say that the eigenvalues are "unimodular.") Demonstrate this using the matrix*

$$\underline{\underline{A}} = \begin{bmatrix} i & 0 \\ 0 & -i \end{bmatrix}$$

by showing that it is unitary and then finding its eigenvalues.

6.18 *Consider two Hermitian matrices* $\underline{\underline{A}}$ *and* $\underline{\underline{B}}$*. Prove both of the following assertions:*

(a) *If* $\underline{\underline{A}}$ *and* $\underline{\underline{B}}$ *commute, that is if* $\underline{\underline{A}}\,\underline{\underline{B}} = \underline{\underline{B}}\,\underline{\underline{A}}$, *then the two matrices share a common set of eigenvectors, albeit with (in principle) different eigenvalues. (You can assume that there is a unique set of eigenvectors for any particular Hermitian matrix.)*

(b) *If* $\underline{\underline{A}}$ *and* $\underline{\underline{B}}$ *share a common set of eigenvectors, then they commute. (Remember that any vector can be written as a linear combination of the eigenvectors of any particular Hermitian matrix.)*

This theorem is critically important for quantum mechanics and the concept of simultaneous measurement.

6.19 *An "ellipsoid" is a three-dimensional surface with three orthogonal symmetry axes, and which appears as an ellipse when viewed along any one axis. Show that the surface described by the points* (x, y, z) *that satisfy*

$$5x^2 + 11y^2 + 5z^2 - 10yz + 2xz - 10xy = 4$$

is an ellipsoid. Find the directions of the axes of symmetry. Also, determine the lengths of the symmetry axes. (This exercise is taken from Riley, Hobson, and Spence Problem 8.26.)

6.20 *The following equation describes a "tilted" ellipsoid in three-dimensional space:*

$$17x^2 - 4xy - 10xz + 20y^2 + 4yz + 17z^2 = 72$$

Plot the ellipsoid in 3D using some computer application. Also plot the symmetry axes on top of the ellipsoid so that it is clear that you have correctly determined the axes. Your result should look something like the following:

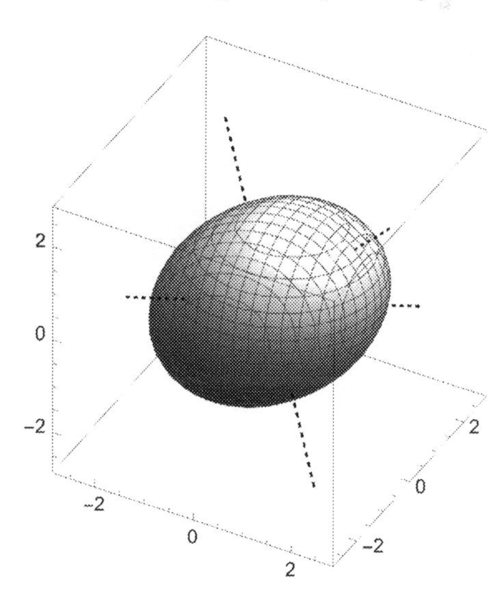

6.21 *Perform the reverse of the previous two problems by first writing down to mutually orthogonal vectors in the three dimensions, and then using the cross product*

to find a third mutually orthogonal vector. Use these three vectors to form a three-dimensional rotation matrix, which you can then apply to an ellipsoid with no cross terms. Then, analyze the rotated ellipsoid to show that you end up with the vectors and ellipsoid that you started with.

6.22 *Find the four eigenfrequencies in terms of $\omega_0^2 \equiv k/m$, and describe the amplitudes for the normal modes to which they correspond, for the four masses connected by five springs on a frictionless horizontal surface, as shown below:*

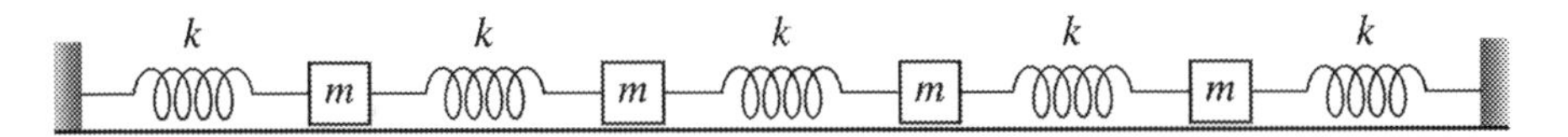

6.23 *For the system shown in Problem 6.22, find and plot the motions of each of the four masses as function of time, when all masses start from rest, with initial positions corresponding to each of the four normal modes. Most of the work for this problem is setting it up correctly in* MATHEMATICA, *identifying each eigenvector component with the correct mass and frequency.*

6.24 *A Lorentz transformation tells you how to convert space and time between two reference frames, call them the "primed" and "unprimed" frames, moving at a velocity v relative to each other, in accordance with the framework of Special Relativity. For a reference frame moving in the x-direction with respect to another frame, the Lorentz transformation is*

$$x' = \gamma(x - vt) \qquad \text{and} \qquad t' = \gamma\left(t - \frac{vx}{c^2}\right)$$

where $\gamma = 1/\sqrt{1-\beta^2}$ and $\beta = v/c$, with c being the speed of light. Then define a vector

$$\underline{x} = \begin{bmatrix} ct \\ x \end{bmatrix}$$

- *(a) Show that the transformation maintains the value of $s^2 = (ct)^2 - x^2$. (We say that the Lorentz transformation maintains the norm of a vector with a "Minkowski metric," instead of a "Euclidean metric.")*
- *(b) Find the Lorentz transformation matrix $\underline{\underline{\Lambda}}$ which takes you from the unprimed frame to the primed frame, by acting on $\underline{x}$.*
- *(c) Write $\underline{\underline{\Lambda}}$ in terms of a single parameter η which combines γ and β. Compare this to a rotation matrix in two dimensions.*
- *(d) Show that the inverse transformation $\underline{\underline{\Lambda}}^{-1}$ corresponds to $v \to -v$ or, equivalently, to the change η to $-\eta$.*

7 Calculus of Variations

Here's a general mathematical problem that turns out to be very important in lots of scientific fields. Imagine you are looking to find some function $f(x)$ that is defined over a range $a \leq x \leq b$. You don't know much about the function, except that you have its values $f(a)$ and $f(b)$, and that there's an integral over this range that involves this function, and you want that integral to be a minimum. In other words, you want to find the function $f(x)$ so that the integral

$$S = \int_a^b F\left(f(x), f'(x), x\right) dx \tag{7.1}$$

is minimized, where $f(a)$ and $f(b)$ have fixed values. You are given the function F of $f(x)$ and its derivatives (and perhaps x as well). How would you go about finding $f(x)$?

The solution to this general problem leads us into the *Calculus of Variations*. The first place you will encounter this problem in a physics course will probably be when you study Lagrangian dynamics in classical mechanics. In this chapter, though, you will see that it is much more useful in general, and we'll do some examples before we're done. There are also some fundamental physics implications, because lots of basic physical laws come down to minimizing the "action" over some "path" through space and time.

You might be interested to know that although it took Einstein a decade to come up with the correct equations for General Relativity, the mathematician David Hilbert figured out how to do it with an action principle and came close to beating Einstein to the punch.[1]

Sometimes the integral S in (7.1) is written as $S[f(x)]$. That is, S takes on a different value for a different function $f(x)$. We refer to $S[f(x)]$ by saying the S is a *functional* of $f(x)$.

We'll start this chapter by doing a specific example, namely showing that the shortest distance between two points is a straight line. Then we'll generalize the technique and do some examples.

7.1 THE SHORTEST DISTANCE BETWEEN TWO POINTS

What curve has the shortest distance between two points in a plane? Obviously, the answer is a straight line, but how would you go about proving that? Well, if you put the two points in the (x, y) plane and assume they are joined by a function $y = f(x)$, then you would integrate the length along this curve and look for the $f(x)$ that gives you the smallest value.

[1] See the article "A comment on the relations between Einstein and Hilbert," Heinrich Medicus, Am. J. Phys. 52(1984)206.

DOI: 10.1201/9781003355656-7

The length of an infinitesimal area element in the (x,y) plane is

$$ds = \sqrt{dx^2 + dy^2}$$

Therefore, the length S between two points a and b in the (x,y) plane is

$$S = \int_a^b ds = \int_a^b \sqrt{dx^2 + dy^2} = \int_a^b \sqrt{1 + \left(\frac{dy}{dx}\right)^2} dx = \int_a^b \sqrt{1 + (f'(x))^2} dx \quad (7.2)$$

and we aim to find the function $y = f(x)$ that minimizes $S[f(x)]$. If we write

$$Y(x) = f(x) + \varepsilon\eta(x)$$

where ε is just some parameter, $f(x)$ is the right function, and $\eta(x)$ is some arbitrary function, then maybe we can find $f(x)$ by considering $S[Y(x)] = S(\varepsilon)$ and setting $dS/d\varepsilon = 0$ when $\varepsilon = 0$. Following our nose and using integration by parts, we have

$$\begin{aligned}
\frac{d}{d\varepsilon} S(\varepsilon) &= \frac{d}{d\varepsilon} \int_a^b \sqrt{1 + (Y'(x))^2} dx = \int_a^b \frac{Y'(x)}{\sqrt{1 + (Y'(x))^2}} \frac{dY'}{d\varepsilon} dx \\
\left.\frac{d}{d\varepsilon} S(\varepsilon)\right|_{\varepsilon=0} &= \int_a^b \frac{f'(x)}{\sqrt{1 + (f'(x))^2}} \eta'(x)\, dx \\
&= \left.\frac{f'(x)}{\sqrt{1 + (f'(x))^2}} \eta(x)\right|_a^b - \int_a^b \frac{d}{dx}\left[\frac{f'(x)}{\sqrt{1 + (f'(x))^2}}\right] \eta(x)\, dx = 0
\end{aligned}$$

Now $\eta(x)$ is an arbitrary function, except that we require $\eta(a) = \eta(b) = 0$ so that $Y(x)$ has the right values at $x = a$ and $x = b$. So, the first term above is zero, and we are left with

$$\int_a^b \frac{d}{dx}\left[\frac{f'(x)}{\sqrt{1 + (f'(x))^2}}\right] \eta(x)\, dx = 0$$

This is in fact an interesting result. It says that the integral is zero no matter what function I choose for $\eta(x)$. The only way to achieve this is to have the factor multiplying $\eta(x)$ be zero. That is

$$\frac{d}{dx}\left[\frac{f'(x)}{\sqrt{1 + (f'(x))^2}}\right] = 0 \quad \text{or} \quad \frac{f'(x)}{\sqrt{1 + (f'(x))^2}} = \text{constant} \quad (7.3)$$

which is of course equivalent to $f'(x)$ =constant. In other words, the curve that connects $x = a$ to $x = b$ over the shortest distance is the one a constant slope, which is a straight line. We have proved our assertion.

7.2 THE EULER-LAGRANGE EQUATIONS

The approach used to find the path with the shortest distance between two points, used in Section 7.1, can easily be generalized to find a condition on the function $F[f(x), f'(x), x]$ in (7.1) that minimizes the functional $S[f(x)]$.

We start the same way, and write

$$Y(x) = f(x) + \varepsilon\eta(x) \qquad \text{with} \qquad \eta(a) = 0 = \eta(b)$$

where $f(x)$ is the correct answer and $\eta(x)$ is otherwise arbitrary. Whereas in Section 7.1 the integrand only depended on $f'(x)$, this time we have the integrand depending, in principle on $f(x)$ as well. Therefore, using the chain rule for a multivariable function,

$$\frac{d}{d\varepsilon}S(\varepsilon) = \int_a^b \left[\frac{\partial F}{\partial Y}\frac{\partial Y}{\partial \varepsilon} + \frac{\partial F}{\partial Y'}\frac{\partial Y'}{\partial \varepsilon}\right] dx$$

and the condition that $f(x)$ minimize $S(0)$ becomes

$$\left.\frac{d}{d\varepsilon}S(\varepsilon)\right|_{\varepsilon=0} = \int_a^b \left[\frac{\partial F}{\partial f}\eta(x) + \frac{\partial F}{\partial f'}\eta'(x)\right] dx = 0$$

Just as in Section 7.1, we treat the second term using integration by parts, so

$$\int_a^b \frac{\partial F}{\partial f'}\eta'(x)dx = \left[\frac{\partial F}{\partial f'}\eta(x)\right]_a^b - \int_a^b \frac{d}{dx}\left(\frac{\partial F}{\partial f'}\right)\eta(x)dx$$

The first term is zero because $\eta(a) = 0 = \eta(b)$, so our minimization condition becomes

$$\int_a^b \left[\frac{\partial F}{\partial f} - \frac{d}{dx}\frac{\partial F}{\partial f'}\right]\eta(x)\,dx = 0$$

Once again, since $\eta(x)$ is arbitrary, we can only satisfy this if the expression that multiplies $\eta(x)$, in square brackets, also zero. That is

$$\frac{\partial F}{\partial f} - \frac{d}{dx}\frac{\partial F}{\partial f'} = 0 \tag{7.4}$$

This is called the *Euler-Lagrange Equation*. Given $F(f(x), f'(x), x)$, it is a second-order differential equation that we solve to find the function $f(x)$ that minimizes $S[f(x)]$.

Let's try it out for the straight line in Section 7.1. In that case

$$F(f(x), f'(x), x) = \sqrt{1 + (f'(x))^2} \tag{7.5}$$

Carrying out the calculations in (7.4) one by one, we have

$$\begin{aligned}
\frac{\partial F}{\partial f} &= 0 \\
\frac{\partial F}{\partial f'} &= \frac{f'(x)}{\sqrt{1+(f'(x))^2}} \\
\frac{d}{dx}\frac{\partial F}{\partial f'} &= \frac{f''(x)}{\sqrt{1+(f'(x))^2}} - \frac{[f'(x)]^2 f''(x)}{\left[1+(f'(x))^2\right]^{3/2}}
\end{aligned}$$

Therefore, (7.4) becomes

$$\begin{aligned}
& -\frac{f''(x)}{\left[1+(f'(x))^2\right]^{1/2}} + \frac{(f'(x))^2 f''(x)}{\left[1+(f'(x))^2\right]^{3/2}} \\
= & -\frac{f''(x)}{\left[1+(f'(x))^2\right]^{3/2}}\left[1+(f'(x))^2-(f'(x))^2\right] \\
= & -\frac{f''(x)}{\left[1+(f'(x))^2\right]^{3/2}} = 0
\end{aligned}$$

which implies that

$$f''(x) = 0$$

In other words, $f(x) = mx + c$ for some constants m and c, a straight line.

7.2.1 IMPORTANT SPECIAL CASES

There are two cases to mention, in which the Euler Lagrange equations reduce to something that is often much simpler to solve than (7.4). In each these cases, this is because (7.4) can be immediately integrated once, depending on the form of F, so we only need to solve a first-order differential equation, instead of second order.

One is the case when $F(f(x), f'(x), x) = F(f'(x), x)$, that is F does not depend explicitly on f. Notice that in Section 7.1 we ended up showing that $f'(x)$ was a constant via (7.3), implying a straight line, whereas above, we showed instead that $f''(x) = 0$. Of course, both are the same, but why did we end up at different places when it seemed like we used the same approach?

The reason is because the functional (7.5) for the distance between two points does not depend on $f(x)$, but only on its derivative. That is $\partial F/\partial f = 0$ so (7.4) implies that

$$\frac{d}{dx}\frac{\partial F}{\partial f'} = 0 \quad \text{so} \quad \frac{\partial F}{\partial f'} = \text{constant} \quad \text{when} \quad F\left(f(x), f'(x), x\right) = F\left(f'(x), x\right) \tag{7.6}$$

which is precisely the statement we concluded with in (7.3).

The second case is $F(f(x), f'(x), x) = F(f(x), f'(x))$, that is F does not depend explicitly on x. To exploit this, we first multiply the Euler-Lagrange equation (7.4) by $f'(x)$ to get

$$f'\frac{\partial F}{\partial f} - f'\frac{d}{dx}\frac{\partial F}{\partial f'} = 0$$

Next we realize that

$$\frac{d}{dx}\left(f'\frac{\partial F}{\partial f'}\right) = f''\frac{\partial F}{\partial f'} + f'\frac{d}{dx}\frac{\partial F}{\partial f'}$$

and also

$$\frac{dF}{dx} = \frac{\partial F}{\partial f}\frac{df}{dx} + \frac{\partial F}{\partial f'}\frac{df'}{dx} + \frac{\partial F}{\partial x} = f'\frac{\partial F}{\partial f} + f''\frac{\partial F}{\partial f'} + \frac{\partial F}{\partial x}$$

Therefore, the Euler Lagrange equation (7.6) becomes

$$f'\frac{\partial F}{\partial f} + f''\frac{\partial F}{\partial f'} - \frac{d}{dx}\left(f'\frac{\partial F}{\partial f'}\right) = \frac{d}{dx}\left(F - f'\frac{\partial F}{\partial f'}\right) - \frac{\partial F}{\partial x} = 0$$

However, the statement that F does not depend explicitly on x means that $\partial F/\partial x = 0$. This all means that

$$F - f'\frac{\partial F}{\partial f'} = \text{constant} \qquad \text{when} \qquad F\left(f(x), f'(x), x\right) = F\left(f(x), f'(x)\right) \tag{7.7}$$

We saw (7.6) applied to the shortest distance between two points, and we will apply (7.7) in Section 7.3. In classical mechanics, you will learn that (7.6) has to do with "conserved quantities," and (7.7) will be used to prove the conservation of energy.

7.2.2 VARIATIONAL NOTATION

A common notation is used which makes it much easier to work with functionals like $S[f(x)]$. If we write $\delta f = \varepsilon\eta(x)$, we can interpret δf as a "small change in $f(x)$" over the range of x that we care about. In that sense,

$$\delta S = S[f(x) + \delta f(x)] - S[f(x)] = 0$$

is equivalent to finding a function $f(x)$ which minimizes S. We say that $S[f(x)]$ is *stationary* when $f(x)$ minimizes the functional. In terms of the explicit form (7.1) we have

$$\delta S = \delta \int_a^b F\left[f(x), f'(x), x\right] dx = \int_a^b \left[\frac{\partial F}{\partial f}\delta f + \frac{\partial F}{\partial f'}\delta f'\right] dx = 0$$

This makes it simple to write down the derivation of (7.4), with the manipulation of the δ's looking just like manipulations of differentials. That is

$$\int_a^b \left[\frac{\partial F}{\partial f}\delta f + \frac{\partial F}{\partial f'}\delta f'\right] dx = \left.\frac{\partial F}{\partial f'}\delta f\right|_a^b + \int_a^b \left[\frac{\partial F}{\partial f} - \frac{d}{dx}\frac{\partial F}{\partial f'}\right]\delta f\, dx = 0 \tag{7.8}$$

and we once again make the argument that $\delta f(x)$ is arbitrary, albeit "small." We know that $\delta f = 0$ when $x = a$ and $x = b$, so the first term on the right is zero and the expression in square brackets must be itself zero.

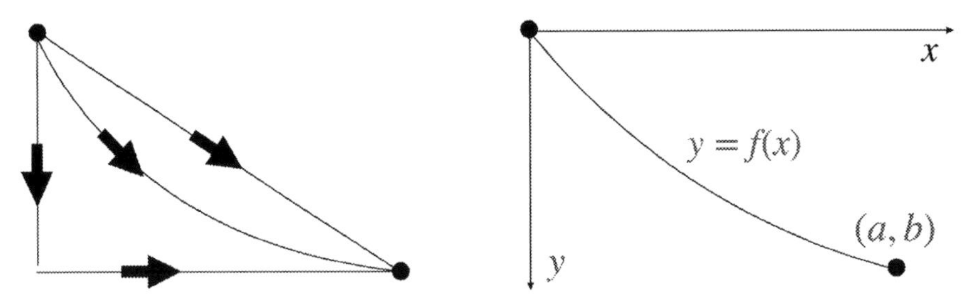

Figure 7.1 The brachistochrone problem is to find the path of shortest time for an object to fall between two points.The left diagram demonstrates that the answer is far from obvious. The right sets up the problem mathematically. Note that positive y is downward.

7.2.3 MORE THAN ONE DEPENDENT VARIABLE

It is straightforward to generalize (7.1) to the case of more than one dependent variable. For example, if we have two dependent variables $f(x)$ and $g(x)$, then

$$S[f(x),g(x)] = \int_a^b F\left(f(x),f'(x),g(x),g'(x),x\right)dx \tag{7.9}$$

and we let both of them vary by "small" functions $\delta f(x)$ and $\delta g(x)$. You get

$$\begin{aligned}\delta S &= \int_a^b \left[\frac{\partial F}{\partial f}\delta f + \frac{\partial F}{\partial f'}\delta f' + \frac{\partial F}{\partial g}\delta g + \frac{\partial F}{\partial g'}\delta g'\right]dx \\ &= \int_a^b \left\{\left[\frac{\partial F}{\partial f} - \frac{d}{dx}\frac{\partial F}{\partial f'}\right]\delta f + \left[\frac{\partial F}{\partial g} - \frac{d}{dx}\frac{\partial F}{\partial g'}\right]\delta g\right\}dx = 0\end{aligned}$$

Allowing $f(x)$ and $g(x)$ to vary independently means that each of the expressions in square brackets must be zero. Therefore, there are two separate Euler-Lagrange equations, namely

$$\frac{\partial F}{\partial f} - \frac{d}{dx}\frac{\partial F}{\partial f'} = 0 \quad \text{and} \quad \frac{\partial F}{\partial g} - \frac{d}{dx}\frac{\partial F}{\partial g'} = 0$$

The generalization to more than two dependent variables is obvious.

7.3 EXAMPLE: THE BRACHISTOCHRONE PROBLEM

Let's use this formalism now to attack a practical problem. Imagine that you have a bead of mass m sliding down along a wire with no friction. The wire starts at a point (x_1, y_1) and ends at a point (x_2, y_2), and you want to know the shape of the wire that lets the bead from the start to the end in the least amount of time.

This is known as the *brachistochrone* problem, from the Greek for "shortest time," and the answer is far from obvious. See the left side of Figure 7.1. Your first thought might be to take the shortest path length, that is, the straight line. Or maybe, you want to fall directly down first, to pick up the greatest speed, then move over to the final point. Or maybe it's somewhere in between.

To set up the problem, see the right side of Figure 7.1. Let the bead start out at the origin, and end up at the point $(x, y) = (a, b)$. The path traveled will be $y = f(x)$, and is fixed at the two endpoints. We want to find the function $f(x)$ that minimizes the time to fall under gravity. This is a clear example of a problem to be solved with the calculus of variations.

Our axes are defined with y going down. That is the bead moves in the $+y$ direction. The time dt it takes for the particle to move a distance $d\ell = \sqrt{dx^2 + dy^2}$ is $d\ell/v$ where v is the particle's speed. Since the bead starts from rest at the origin, conservation of energy says

$$\frac{1}{2}mv^2 - mgy = 0 \qquad \text{so} \qquad v = \sqrt{2gy}$$

We can therefore write the time it takes for the bead to travel along the path $y = f(x)$ as

$$T[f(x)] = \int_{0,0}^{a,b} \frac{d\ell}{v} = \int_{0,0}^{a,b} \frac{\sqrt{dx^2 + dy^2}}{\sqrt{2gy}} = \frac{1}{\sqrt{2g}} \int_0^a \sqrt{\frac{1 + (f'(x))^2}{f(x)}}\, dx$$

Finding $f(x)$ that minimizes $T[f(x)]$ means to apply the Euler Lagrange equation (7.4) to

$$F\left(f(x), f'(x), x\right) = \sqrt{\frac{1 + (f'(x))^2}{f(x)}} = \left[\frac{1 + (f'(x))^2}{f(x)}\right]^{1/2}$$

This time $F\left(f(x), f'(x), x\right)$ does not explicitly depend on x, so we can use the Euler Lagrange equation as integrated in (7.7). That is

$$\begin{aligned} F - f'\frac{\partial F}{\partial f'} &= \left[\frac{1 + (f'(x))^2}{f(x)}\right]^{1/2} - \frac{1}{(f(x))^{1/2}} \frac{(f'(x))^2}{\left[1 + (f'(x))^2\right]^{1/2}} \\ &= \frac{1}{(f(x))^{1/2}} \frac{1}{\left[1 + (f'(x))^2\right]^{1/2}} = \text{constant} \equiv \frac{1}{c^{1/2}} \end{aligned}$$

Squaring both sides gives us the differential equation

$$f(x)\left[1 + (f'(x))^2\right] = c$$

where c is a constant that we will determine shortly from the boundary conditions.

If we write $y = f(x)$ the differential equation becomes

$$\frac{dy}{dx} = \sqrt{\frac{c - y}{y}}$$

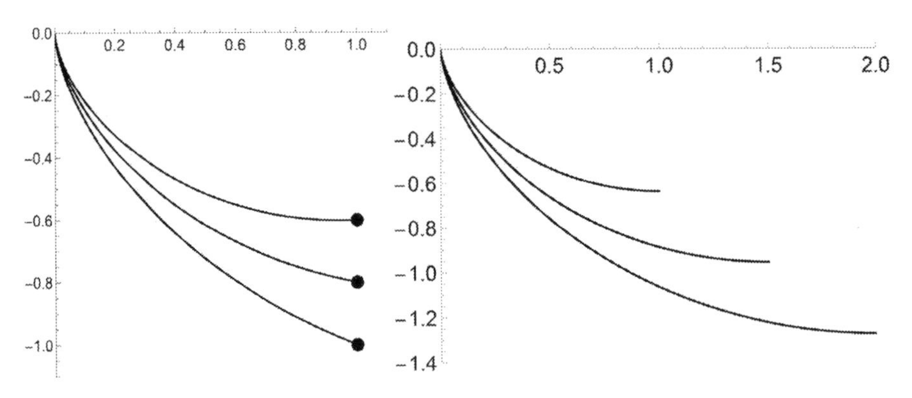

Figure 7.2 Different solutions to the brachistochrone problem. The left is where the endpoint is fixed. The right is when the vertical position of the endpoint is allowed to float.

where we note that, the way we have defined the axes in Figure 7.1, we expect the derivative to be positive. Writing this as

$$\sqrt{\frac{y}{c-y}}\,dy = dx$$

we see that the left side is easy to integrate if we make the substitution $y = c\sin^2\theta$. Then $c - y = c\cos^2\theta$ and $dy = 2c\sin\theta\cos\theta\,d\theta$, and

$$\frac{\sin\theta}{\cos\theta}2c\sin\theta\cos\theta\,d\theta = 2c\sin^2\theta\,d\theta = c(1-\cos 2\theta) = dx$$

Integrating this and putting it together with our substitution for y gives us

$$x = \frac{c}{2}(2\theta - \sin 2\theta) + d \tag{7.10a}$$

$$y = c\sin^2\theta \tag{7.10b}$$

where c and d are constants. These parametric equations describe a curve called a *cycloid*, which is usually described as the path of a point on the rim of a rolling wheel of radius c. It apparently is also the shape of the path that minimizes the travel time between two points for a bead subject to gravity.

To complete the problem, we have to determine c and d. Since $\theta = 0$ gives $y = 0$ and $x = d$, it is clear that $d = 0$ since the curve includes the origin. Therefore (7.10) can then be solved for c and the value of θ that gives $(x,y) = (a,b)$. This typically requires a numerical solution.

Figure 7.2 on the left shows the solution for the endpoints (a,b), each with $a = 1$ and $b = 0.6$, 0.8, and 1.0. (I did not bother to reverse the sign of the vertical axis.)

Interestingly, we can find solutions if we instead provide only the value of a, and let the vertical position of the endpoint float. This would seem to spoil our formalism, where we required that the endpoints be fixed. However, looking back at our derivation of the Euler Lagrange equations in (7.8), we could instead require that $\partial F/\partial f' = 0$ at the endpoint. In our case, this means that

$$\frac{1}{(f(x))^{1/2}} \frac{f'(x)}{\left[1+(f'(x))^2\right]^{1/2}} = 0 \qquad \text{for} \qquad x = a$$

which in turn implies that the slope $y' = 0$ at $x = a$. From (7.10) this implies that

$$c \sin 2\theta = 0 \qquad \text{when} \qquad a = c\theta = c\pi/2$$

Plots with this approach, for $a = 1$, 1.5, and 2.0 are shown on the right in Figure 7.2.

7.4 EXAMPLE WITH CONSTRAINTS: LAGRANGE MULTIPLIERS

It is not uncommon to have a variational calculus problem that involves *constraints* on the quantities involved. There is a general approach to all such problems that involves the use of *Lagrange Multipliers.* We will illustrate the approach here with one type of constraint, namely a different integral over the same independent variable which has to be kept at a fixed value.

Once again, our problem is to find the function $f(x)$ which minimizes the functional

$$S = \int_a^b F\left[f(x), f'(x), x\right] dx \tag{7.11}$$

This time, however, there is a constraint that the quantity

$$L = \int_a^b G\left[f(x), f'(x), x\right] dx \tag{7.12}$$

must be kept constant. For example, L might represent the length of the curve $y = f(x)$. Keeping L constant is the same as writing $\delta L = 0$ as we vary $\delta f(x)$. Since $\delta S = 0$ for the correct $f(x)$, we can get a modified form of the Euler-Lagrange equation by writing

$$\delta(S + \lambda L) = \delta \int_a^b \left\{F\left[f(x), f'(x), x\right] + \lambda G\left[f(x), f'(x), x\right]\right\} dx = 0 \tag{7.13}$$

for some constant λ. This modified Euler-Lagrange equation will automatically include the constraint (7.12). An additional constant λ, known as Lagrange multiplier, will be included in the differential equation for $f(x)$ that results, but it can be determined using the constraint equation.

You will see this in an advanced course in classical mechanics, where the Lagrange multipliers turn out to be the "forces" which lead to constrained motion of a system. See, for example, Section 19 of "Theoretical Mechanics of Particles and Continua" by Fetter and Walecka.

Let's again illustrate this approach with a specific problem. Figure 7.3 shows a "rope" of fixed length L is attached to the x-axis between the points $x = a$ and $x = b$.

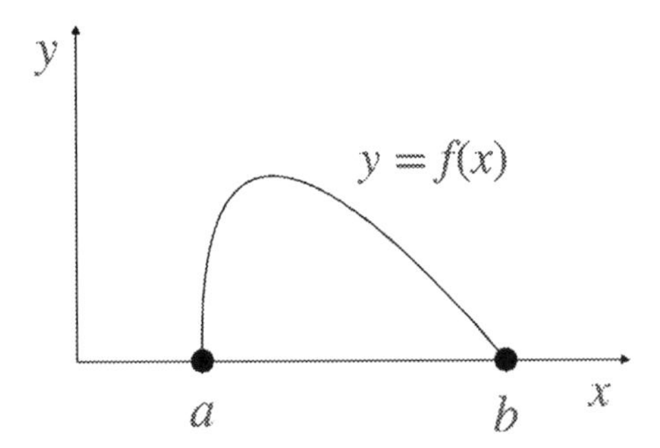

Figure 7.3 An example of a calculus of variations problem with an integral constraint. The goal is to find the function $y = f(x)$ which is fixed to the x-axis at $x = a$ and $x = b$ and which encloses the maximum area underneath it, subject to the constraint that the length $L > b - a$ of the curve is fixed. You can think of the curve as a fixed length of rope.

The job is to find the shape of the rope which maximizes the area it encloses over the x-axis. We are therefore looking to maximize the functional

$$S[f(x)] = \int_a^b f(x)\,dx$$

subject to the constraint that

$$L = \int_a^b ds = \int_a^b \left[1 + (f'(x))^2\right]^{1/2} dx$$

remains fixed. The modified Euler-Lagrange equation is derived from

$$\begin{aligned}
\delta(S + \lambda L) &= \delta \int_a^b \left\{ f(x) + \lambda \left[1 + (f'(x))^2\right]^{1/2} \right\} dx \\
&= \int_a^b \left\{ \delta f(x) + \lambda \frac{f'(x)}{\left[1 + (f'(x))^2\right]^{1/2}} \delta f'(x) \right\} dx \\
&= \lambda \frac{f'(x)}{\left[1 + (f'(x))^2\right]^{1/2}} \delta f(x) \Bigg|_a^b \\
&\quad + \int_a^b \left\{ 1 - \lambda \frac{d}{dx} \frac{f'(x)}{\left[1 + (f'(x))^2\right]^{1/2}} \right\} \delta f(x)\,dx = 0
\end{aligned}$$

where we once again make use of integration by parts. Now, as we have observed previously, the first term on the right hand size is zero because $\delta f(a) = \delta f(b) = 0$. That is, the endpoints of the rope are fixed to the x-axis. Also as before, $\delta f(x)$ is arbitrary, so to have the integral equal to zero implies that

$$1-\lambda\frac{d}{dx}\frac{f'(x)}{\left[1+(f'(x))^2\right]^{1/2}}=0$$

This is the differential equation we need to solve in order to come up with the function $y=f(x)$. This form is easy to integrate once, and we get

$$\frac{\lambda f'(x)}{\left[1+(f'(x))^2\right]^{1/2}}=x+c$$

for some constant c. Squaring both sides and writing $dy/dx=f'(x)$ we get the differential equation

$$\lambda^2\left(\frac{dy}{dx}\right)^2=(x+c)^2\left[1+\left(\frac{dy}{dx}\right)^2\right] \qquad \text{so} \qquad dy=\frac{x+c}{[\lambda^2-(x+c)^2]^{1/2}}dx$$

The right side is not hard to integrate if we make the substitution

$$x+c=\lambda\sin\theta \qquad \text{so} \qquad dx=\lambda\cos\theta\, d\theta$$

in which case the differential equation becomes

$$dy=\frac{\lambda\sin\theta}{\lambda\cos\theta}\lambda\cos\theta\, d\theta=\lambda\sin\theta\, d\theta \qquad \text{so} \qquad y+c'=-\lambda\cos\theta$$

where c' is some other constant. It is simple to eliminate the variable θ, and we get

$$(x+c)^2+(y+c')^2=\lambda^2$$

The answer is a circular arc, probably what you would have guessed. The Lagrange multiplier λ has an obvious physical interpretation as the radius of the circle. In order to find c and c', we need to solve

$$\begin{aligned} (a+c)^2+(c')^2 &= \lambda^2 \\ \text{and} \quad (b+c)^2+(c')^2 &= \lambda^2 \end{aligned}$$

for c and c', in terms of a, b, and λ. Subtracting these two equations gives $a+c=\pm(b+c)$, but the plus sign gives nonsense, so $c=(a+b)/2$. The result is simpler if we translate the x-axis so that $a=-x_0$ and $b=x_0$ which means that $c=0$ and $c'=\pm(\lambda^2-x_0^2)^{1/2}$. See Figure 7.4. Clearly, as you would have expected, the radius λ of the circle has to be at least as large as the displacement of the fixed points from the origin.

We see that the center of the circle is along the y axis, at a distance c' above (or below) the x-axis. If θ represents the angular variable that traces the circle, then

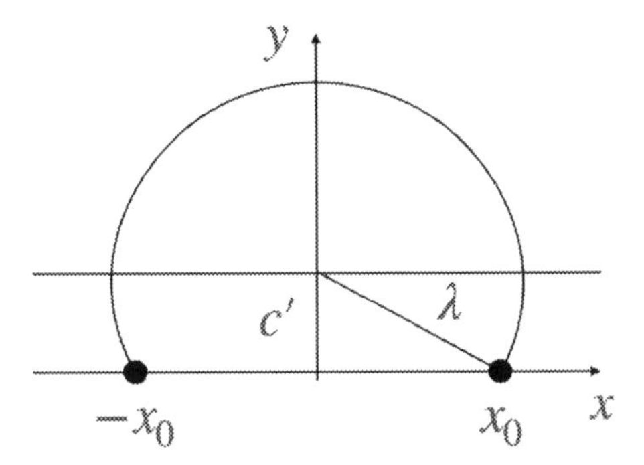

Figure 7.4 Solution to the problem of maximizing the area contained inside a curve made from a fixed length of rope. The shape is a circle and the Lagrange multiplier λ is clearly interpreted as the radius of the circle. The vertical displacement c' of the circle's center is determined by the positions of the endpoints along with the radius.

$\tan\theta = \pm x_0/c'$ gives the range of angles $\Delta\theta = 2\tan^{-1}(x_0/c')$ that are excluded from the circle. The length constraint then becomes

$$L = (2\pi - \Delta\theta)\lambda$$

However, it is more important to realize the physical interpretation of λ, namely the radius of the circle. Sure, if we want to fully solve the problem to find the shape of the curve that maximizes the area for these fixed points and length, this is what we need. However, the "physics" is more in the interpretation, and that's usually what's really important.

There's one other question, which in fact bothers me because I don't know how to answer it. We wrote our functional as "the area under a curve $y = f(x)$" which apparently implies that $f(x)$ is a single valued function that sits above the x-axis. However, we've come up with a solution that isn't (necessarily) single valued if the rope is long enough, and that can also be below the axis. Perhaps understanding "below the axis" isn't so hard – it's the minimum, not the maximum – and maybe the fact that we were careless with signs when squaring things and taking square roots explains the lack of single-valuedness. Nevertheless, I thought I would mention it.

EXERCISES

7.1 *A particle of mass m moves in one dimension x over some time interval $t_1 \leq t \leq t_2$, under the influence of a force $F(x) = -dV/dx$, some function $V(x)$. Show that finding the function $x(t)$ which minimizes*

$$S = \int_{t_1}^{t_2} L(x,\dot{x})\,dt \qquad \text{where} \qquad L(x,\dot{x}) = \frac{1}{2}m\dot{x}^2 - V(x)$$

is the same as showing that $x(t)$ is determined by Newton's Second Law of Motion. Then show that since L does not explicitly depend on t, total mechanical energy is conserved.

7.2 *The text discusses two special cases which allowed the Euler-Lagrange equations to be integrated once. In fact, when we first started talking about the Calculus of Variations, we found that a straight line gave the shortest distance between two points, using the first of these special cases. Show that the functional for the shortest path between two points is also an example of the second special case, and use that form to show that solution is a straight line.*

7.3 *Find and plot the brachistochrone solution for a bead starting at the origin and ending at* $(a,b) = (1,2)$.

7.4 *A "surface of revolution" is formed when a shape given by* $y = f(x)$ *for* $a \le x \le b$ *is rotated about the x-axis, as shown in the figure below:*

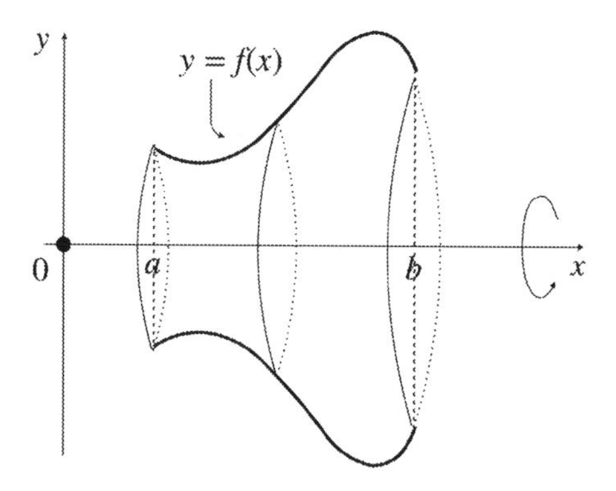

Find the form of the function $f(x)$ *which minimizes the surface area. You don't need to solve for the constants of integration in terms of the fixed points of* $f(x)$ *at* $x = a$ *and* $x = b$.

7.5 *Two horizontal identical circular hoops, each having radius R, are coaxial and separated vertically by a distance 2h. A continuous soap film is attached to the hoops and drapes between them. Assuming that the surface tension of the film is proportional to its surface area, and that the film is in equilibrium when the surface tension is minimized, find the shape of the soap film. You can ignore the mass of the film, and you can leave your answer in terms of a single undetermined constant.*

7.6 *A point particle of mass m moves in space as a function of time, following the position vector* $\vec{r}(t) = \hat{\imath}x(t) + \hat{\jmath}y(t) + \hat{k}z(t)$. *Assuming the particle moves according to the Principle of Least Action, that is, the path* $\vec{r}(t)$ *is the one that minimizes the functional*

$$S[\vec{r}(t)] = \int_{t_1}^{t_2} L(x,y,z,\dot{x},\dot{y},\dot{z})\,dt \qquad \text{where} \qquad L = \frac{1}{2}m\dot{\vec{r}}^2 - V(\vec{r})$$

show that $\vec{F} = m\vec{a}$, *aka "the equation of motion," where* $\vec{F} = -\vec{\nabla}V$ *and* $\vec{a} = \ddot{\vec{r}}$.

7.7 *Two identical masses m slide on a frictionless horizontal surface and are each connected to a fixed outside wall and to each other by identical springs of stiffness k. The positions of the masses are given by $x_1(t)$ and $x_2(t)$. Knowing that the potential energy of a spring that is compressed or stretched a distance Δ is $k\Delta^2/2$, find the equation of motion for each of the two masses using the Principle of Least Action and the Euler-Lagrange equation. Check your answer against the example we have studied in class.*

7.8 *A particle of mass m moves horizontally on a frictionless surface defined by the x,y plane. Convert to polar coordinates ρ and ϕ, and find the Euler-Lagrange equations of motion. Assuming that potential energy $V = V(\rho)$, that is it has no ϕ-dependence, show that one of the equations of motion leads to a "conserved quantity," that is, something that does not change with time. What is the common name for this conserved quantity?*

7.9 *A highly flexible cable of linear mass density μ and fixed length ℓ hangs motionless in the vertical plane, where its shape minimizes the gravitational potential energy. The cable is fixed at two points at the same vertical position, but separated horizontally by a distance $d < \ell$. Assuming the shape of the cable is given by the function $f(x)$ where x measures the horizontal position, write the integrals that express (a) the gravitational potential energy and (b) the total length of the cable. Combine these integrals and use this to derive a constrained Euler-Lagrange equation that can be solved to find the shape of the hanging cable. Solve this equation for the shape. You don't need to get the result in terms of ℓ and d, but show how you would do that, in principle.*

7.10 *A chain of length $L > 2a$ hangs freely between two points $x = \pm a$ on the x-axis in the xy plane. Find the equation $f(x)$ that describes the resting shape of the chain, assuming that this is the shape that minimizes the center of gravity. Of course, the length L must remain fixed, and $f(\pm a) = 0$. Eliminate whichever constants of integration are easiest, but come up with a simple physical interpretation of the Lagrange multiplier used to set the length constraint.*

8 Functions of a Complex Variable

We have been rather cavalier about complex variables throughout this book so far. After mentioning them only briefly in Section 1.1.1.1, we made use of them from time to time when it was handy. Probably most useful was the relationship

$$z = x + iy = re^{i\phi} = r\cos\phi + i\,r\sin\phi$$

which showed how to represent a complex number z in terms of two real numbers x and y which identified a point in the complex plane with Cartesian coordinates, or in terms of two different real numbers r and ϕ which identify the same point but in plane polar coordinates. The connection was made using Euler's Formula, which we "derived" in Section 2.4.

This chapter will get much more serious about complex variables, in particular by discussing the theory and applications of *functions* of complex variables. These functions are mappings $\mathbb{C} \mapsto \mathbb{C}$, which we can write generically as

$$w = f(z) = u(x,y) + iv(x,y) \qquad \text{where} \qquad z = x + iy \tag{8.1}$$

with $x, y, u, v \in \mathbb{R}$ and $z, w \in \mathbb{C}$. In fact, it will be useful to realize that these functions are also mappings $\mathbb{R}^2 \mapsto \mathbb{R}^2$, which means they should have properties of functions in two spatial dimensions. Indeed, we'll borrow some things we proved about vector calculus in Chapter 4.

8.1 DIFFERENTIABILITY AND CONVERGENCE

It can seem that the generalization to complex functions from real functions is simple, and not worth a lot of discussion, but in fact that is not the case. I will illustrate this with a couple of examples of fundamental differences between real and complex functions.

For our first example, consider the real function

$$f(x) = 2\int_0^x |x'|\,dx' = \begin{cases} +x^2 & x \geq 0 \\ -x^2 & x \leq 0 \end{cases} \tag{8.2}$$

which is plotted in Figure 8.1. It looks like a perfectly reasonable continuous function passing through the origin. In fact, it has a perfectly definable derivative at $x = 0$. That is, taking ε to be positive,

$$f'(0) = \lim_{\varepsilon\to 0} \frac{f(+\varepsilon) - f(-\varepsilon)}{2\varepsilon} = \lim_{\varepsilon\to 0} \frac{(+\varepsilon)^2 - (-(-\varepsilon)^2)}{2\varepsilon} = \lim_{\varepsilon\to 0} \frac{2\varepsilon^2}{2\varepsilon} = \lim_{\varepsilon\to 0} \varepsilon = 0$$

DOI: 10.1201/9781003355656-8

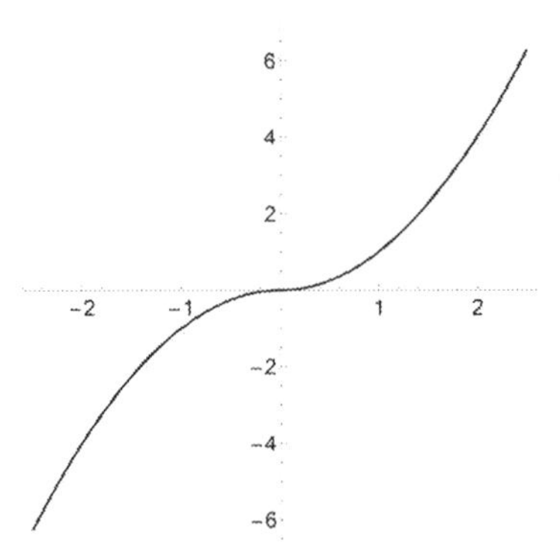

Figure 8.1 A plot of the (real) function given by (8.2). This is a smooth, that is continuous, function, passing through the point $(x,y) = (0,0)$. It also has a well-defined derivative, equal to zero, at $x = 0$. However, higher derivatives do not exist at $x = 0$. We will see that you can never have this situation with a complex function. That is, if it is differentiable once, then it it is differentiable an infinite number of times.

What's more, the first derivative is continuous at $x = 0$, since $f(x) = 2x$ for $x > 0$ and $f(x) = -2x$ when $x < 0$, and both of these approach $f'(0)$ for $x \to 0$, regardless of whether I approach from the negative of positive direction.

However, there is a problem with the second derivative. For $x > 0$, it is clear that $f''(x) = +2$, but for $x < 0$, $f''(x) = -2$. There is a clear discontinuity in the second derivative at $x = 0$. That is, $f''(0)$ does not exist, even though the function and its first derivative both exist and are continuous.

We will learn that this never happens for functions of a complex variable. If a function is differentiable once, then it will be differentiable for an infinite number of times. This is due to the property of *analyticity*, which we will cover in Section 8.3. The catch is that in order for a complex function to be analytic, it needs to satisfy a particularly stringent condition.

Now a second example. Consider the (real) function

$$f(x) = \frac{1}{1+x^2}$$

This is perfectly well-defined for all real numbers. However, if we perform a Taylor expansion about $x = 0$ we find

$$f(x) = 1 - x^2 + x^4 - x^6 + \cdots$$

and we are going to get into trouble for $|x| \geq 1$. We might have expected something like this if the function were $1/(1-x^2)$ because it would be ill defined at $x = \pm 1$. The problem here is that if we made the function complex, that is

$$f(z) = \frac{1}{1+z^2}$$

then it would be ill defined for $z = \pm i$. We therefore say that this function, as well as the function $1/(1-x^2)$, have *radii of convergence* given by $|z| < 1$. Obviously, the idea of functions of real numbers is somehow incomplete, and we need to consider what happens when we extend the function into the complex plane.

We will see that it becomes natural to think of functions of real variables as a special case of functions of complex variables. In fact, we will find that doing things like integrating real functions is better thought of in terms of integrals over some path in the complex plane.

First, though, let's do some preliminary and simple investigations of the properties of complex functions.

8.2 EXAMPLES OF COMPLEX FUNCTIONS

In this section we will go through various specific examples of complex functions written in the form (8.1). We'll refer back to these examples when we study more fundamental properties of analytic functions.

8.2.1 POWER LAWS

First consider $w = f(z) = z^2$ where $z = x + iy$. It is simple to put this in the form (8.1). We naturally write

$$w = z^2 = (x+iy)^2 = x^2 - y^2 + 2ixy = u(x,y) + iv(x,y) \tag{8.3a}$$

where

$$u(x,y) = x^2 - y^2 \qquad \text{and} \qquad v(x,y) = 2xy \tag{8.3b}$$

We can think about the derivative of $w = f(z)$ with respect to z. We sort of expect the answer to be $f'(z) = 2z$, but can we prove it? We would proceed to write

$$f'(z) = \frac{dw}{dz} = \lim_{\Delta z \to 0} \frac{f(z+\Delta z) - f(z)}{\Delta z} = \lim_{\Delta z \to 0} \frac{(z+\Delta z)^2 - z^2}{\Delta z} = \lim_{\Delta z \to 0} \frac{2z\Delta z + (\Delta z)^2}{\Delta z}$$

but we have to be aware of a potential complication, because $\Delta z = \Delta x + i\Delta y$, so Δz can approach zero in various ways, depending on how you take the limits $\Delta x \to 0$ and $\Delta y \to 0$. Writing this all out, and tossing out higher orders in Δx and Δy, we find

$$f'(z) = \lim_{\Delta x \to 0, \Delta y \to 0} \frac{2(x+iy)(\Delta x + i\Delta y)}{\Delta x + i\Delta y}$$

and now you can see clearly that if we take $\Delta z \to 0$ by, first, putting $\Delta y = 0$ and then taking $\Delta x \to 0$, or, second, putting $\Delta x = 0$ and then taking $\Delta y \to 0$, then in either case we get $f'(z) = 2(x+iy) = 2z$.

In fact, we can be more general and write $\Delta y = a\Delta x$ for some constant a, in which case

$$f'(z) = \lim_{\Delta x \to 0} \frac{2(x+iy)\Delta x(1+ia)}{\Delta x(1+ia)} = 2(x+iy) = 2z$$

so we are confident that the derivative is what we expect. We will investigate the general circumstances under which the derivative of a complex function makes sense in Section 8.3.

It follows directly from this analysis that higher powers of positive integers are also well-defined and differentiable. This means that infinite series of the form

$$f(z) = \sum_{n=0}^{\infty} c_n z^n$$

are also well-defined.

Powers that are not positive integers, however, can be problematic. For example, functions that include terms or factors like

$$f(z) = \frac{1}{z} \qquad \text{or} \qquad f(z) = \frac{1}{z - z_0}$$

are of course singular at $z = 0$ or $z = z_0$, so are not differentiable there. We say that these functions of *poles* at $z = 0$ or $z = z_0$.

Different difficulties arise for non-integer positive powers. For example, consider the function

$$f(z) = z^{1/2}$$

It is easiest to analyze this by expressing z in polar coordinates, that is $z = re^{i\phi}$, so

$$f(z) = r^{1/2} e^{i\phi/2}$$

Now if $\phi \to \phi + 2\pi$, then $z \to re^{i(\phi+2\pi)} = re^{i\phi} e^{2\pi i} = re^{i\phi}$ so z is unchanged. However

$$f(z) \to r^{1/2} e^{i(\phi+2\pi)/2} = r^{1/2} e^{i\phi/2} e^{i\pi} = -r^{1/2} e^{i\phi/2}$$

That is, $f(z)$ changes sign. Therefore, we need to agree on a "standard" range of the phase ϕ, and that is $-\pi/2 < \phi \leq +\pi/2$.

8.2.2 SPECIAL FUNCTIONS

Most of the special functions discussed in Sections 1.5 and 3.6 are more generally expressed as functions of a complex variable $z = x + iy$. It is easiest to start with

$$f(z) = e^z = e^x e^{iy}$$

from which we can define

$$\cos z = \frac{e^{iz} + e^{-iz}}{2} \qquad \text{and} \qquad \sin z = \frac{e^{iz} - e^{-iz}}{2i}$$

$$\text{and} \qquad \cosh z = \frac{e^{z} + e^{-z}}{2} \qquad \text{and} \qquad \sinh z = \frac{e^{z} - e^{-z}}{2}$$

For the logarithm, we naturally write $z = re^{i\phi}$ and so

$$\log z = \log r + i\phi$$

where now it is obvious that we need to stick to the standard range of ϕ.

8.2.3 PECULIAR EXAMPLES

We can also define some oddball functions like

$$f(z) = z^* = x - iy \qquad \text{and} \qquad f(z) = |z| = (x^2 + y^2)^{1/2}$$

Taking derivatives of these sorts of things will turn out to be problematic.

8.3 ANALYTICITY

If a function of a complex variable has a derivative that exists at some point in the complex plane, then we say that the function is analytic at that point. We saw in Section 8.2.1 that we needed to consider approaching from either the x or y direction, or any direction for that matter, when calculating the derivative of the function $f(z)=z^2$. In this section we will formalize this requirement for general functions of a complex variable.

8.3.1 CAUCHY-RIEMANN RELATIONS

Go back to the basics and you'll see a potential problem. If you approach the limit to the point (x,y) in the complex plane along a line of constant y, then

$$\begin{aligned} \frac{dw}{dz} &= \lim_{\varepsilon\to 0}\frac{f(z+\varepsilon)-f(z)}{\varepsilon} \\ &= \lim_{\varepsilon\to 0}\frac{u(x+\varepsilon,y)-u(x,y)}{\varepsilon}+i\lim_{\varepsilon\to 0}\frac{v(x+\varepsilon,y)-v(x,y)}{\varepsilon} \\ &= \frac{\partial u}{\partial x}+i\frac{\partial v}{\partial x} \end{aligned} \tag{8.4}$$

On the other hand, if you approach along a line of constant x, then

$$\begin{aligned} \frac{dw}{dz} &= \lim_{\varepsilon\to 0}\frac{f(z+i\varepsilon)-f(z)}{i\varepsilon} \\ &= \frac{1}{i}\lim_{\varepsilon\to 0}\frac{u(x,y+\varepsilon)-u(x,y)}{\varepsilon}+\lim_{\varepsilon\to 0}\frac{v(x,y+\varepsilon)-v(x,y)}{\varepsilon} \\ &= -i\frac{\partial u}{\partial y}+\frac{\partial v}{\partial y} \end{aligned} \tag{8.5}$$

Therefore, in order to consistently define the derivative, the two forms for dw/dz in (8.4) and (8.5) must be equivalent. Therefore, we must have

$$\frac{\partial u}{\partial x} = \frac{\partial v}{\partial y} \tag{8.6a}$$

$$\text{and} \quad \frac{\partial v}{\partial x} = -\frac{\partial u}{\partial y} \tag{8.6b}$$

These are called the *Cauchy-Riemann Relations*. Any function $f(z)$ which obeys these relations is called *analytic*.

Now just because we've come up with conditions based on ε approaching zero along lines of constant x or constant y, doesn't mean that the derivative works for $\varepsilon\to 0$ from any direction. It may seem reasonable because we showed it works for two orthogonal directions, but there are more restrictions that I won't go into. For our purposes, however, we can just assume that a function $f(z)$ is analytic if and only if (8.6) holds.

There is an interesting connection between the Cauchy-Riemann Relations and the form of a complex number in terms of a 2×2 matrix, given in (6.19). If we think of a complex function as a "coordinate transformation" from x and y to u and v, then the differentials are related as

$$\begin{aligned} du &= \frac{\partial u}{\partial x}dx + \frac{\partial u}{\partial y}dy \\ dv &= \frac{\partial v}{\partial x}dx + \frac{\partial v}{\partial y}dy \end{aligned}$$

which can be written as $\underline{dw} = \underline{\underline{J}}\,\underline{dz}$ where

$$\underline{dw} = \begin{bmatrix} du \\ dv \end{bmatrix} \qquad \text{and} \qquad \underline{dz} = \begin{bmatrix} dx \\ dy \end{bmatrix}$$

are representations of the complex differentials as two-dimensional vectors, and

$$\underline{\underline{J}} = \begin{bmatrix} \partial u/\partial x & \partial u/\partial y \\ \partial v/\partial x & \partial v/\partial y \end{bmatrix}$$

is called the *Jacobian matrix*. If the transformation is an analytic function, then

$$\underline{\underline{J}} = \begin{bmatrix} \partial u/\partial x & -\partial v/\partial x \\ \partial v/\partial x & \partial u/\partial x \end{bmatrix}$$

which is exactly the form (6.19). In other words, for an analytic function, the "coordinate transformation" reduces to the product of complex numbers.

Let's do a couple of examples, starting with $f(z) = z^2$. See (8.3). We have

$$\frac{\partial u}{\partial x} = \frac{\partial}{\partial x}(x^2 - y^2) = 2x \qquad \text{and} \qquad \frac{\partial v}{\partial y} = \frac{\partial}{\partial y}(2xy) = 2x$$

so (8.6a) is satisfied. We also have

$$\frac{\partial v}{\partial x} = \frac{\partial}{\partial x}(2xy) = 2y \qquad \text{and} \qquad \frac{\partial u}{\partial y} = \frac{\partial}{\partial y}(x^2 - y^2) = -2y$$

and (8.6b) is also satisfied. Therefore, the function is analytic. In fact, it is analytic for every z in the complex plane. We therefore say that the function is *entire*.

Now consider $f(z) = z^*$ so that $u(x,y) = x$ and $v(x,y) = -y$. This satisfies (8.6b), but violates (8.6a), i.e. "$1 = -1$," everywhere. This function is analytic nowhere in the complex plane.

8.3.2 INFINITE DIFFERENTIABILITY

It is easy to see that if $f(z)$ is analytic at a point z, then $f'(z)$ is also analytic at z. In other words, if $f(z)$ is differentiable once at z, then it is differentiable for an infinite number of times. If we write the derivative of $f(z)$ as

$$f'(z) = \frac{\partial u}{\partial x} + i\frac{\partial v}{\partial x}$$

then we need to demonstrate that (8.6) hold if $u \to \partial u/\partial x$ and $v \to \partial v/\partial x$. Testing (8.6a), we have

$$\frac{\partial}{\partial x}\frac{\partial u}{\partial x} = \frac{\partial}{\partial y}\frac{\partial v}{\partial x} \quad \text{but} \quad \frac{\partial u}{\partial x} = \frac{\partial v}{\partial y}$$

since $f(z)$ is analytic. Therefore, (8.6a) is satisfied, assuming $u(x,y)$ and $v(x,y)$ are well-behaved enough so that the order of the derivatives can be reversed.

Similarly, testing (8.6b) gives

$$\frac{\partial}{\partial x}\frac{\partial v}{\partial x} = -\frac{\partial}{\partial y}\frac{\partial u}{\partial x} \quad \text{but} \quad \frac{\partial v}{\partial x} = -\frac{\partial u}{\partial y}$$

so (8.6b) is also satisfied. Thus, $f'(z)$ is analytic if $f(z)$ is analytic.

8.4 CONTOUR INTEGRATION

It should be clear that if we are going to integrate some function $f(z)$ between z_1 and z_2, then we'll need to specify the path we take in the complex plane to connect these two complex numbers. Consequently, complex integration, generally referred to as *contour integration*, will make use of many of the techniques we developed in Chapter 4, especially Section 4.3.

It is not too hard to turn a contour integral into a set of real integrals. If C is the curve along which we wish to integrate from $z_1 = x_1 + iy_1$ to $z_2 = x_2 + iy_2$, then

$$\begin{aligned}\int_C f(z)\,dz &= \int_C [u(x,y) + iv(x,y)]\,[dx + i\,dy] \\ &= \int_C [u(x,y)\,dx - v(x,y)\,dy] + i\int_C [v(x,y)\,dx + u(x,y)\,dy] \qquad (8.7) \\ &= \int_{t_1}^{t_2} \left[u(x,y)\frac{dx}{dt} - v(x,y)\frac{dy}{dt}\right] dt + i\int_{t_1}^{t_2} \left[v(x,y)\frac{dx}{dt} + u(x,y)\frac{dy}{dt}\right] dt \qquad (8.8)\end{aligned}$$

where the form (8.8) makes use of an assumed parameterization of the curve by the functions $x(t)$ and $y(t)$. This in fact would be a practical way to carry out the integral.

For example, suppose we want to integrate from $(x_1, y_1) = (1,0)$ to $(x_2, y_2) = (-1,0)$ along a semicircle C traced counter clockwise through the upper half of the complex plane. See Figure 8.2. Then $x(t) = \cos t$, $y(t) = \sin t$, $t_1 = 0$, and $t_2 = \pi$. For the function $f(z) = z = x + iy$, we have

$$\begin{aligned}\int_C f(z)\,dz &= \int_0^{\pi} [\cos t(-\sin t) - \sin t(\cos t)]\,dt \\ &\quad + i\int_0^{\pi} [\sin t(-\sin t) + \cos t(\cos t)]\,dt \\ &= \int_0^{\pi} 2\sin t\cos t\,dt + i\int_0^{\pi} \left[\cos^2 t - \sin^2 t\right] dt \\ &= \int_0^{\pi} \sin 2t\,dt + i\int_0^{\pi} \cos 2t\,dt = -\frac{1}{2}\cos 2t\Big|_0^{\pi} + i\,\frac{1}{2}\sin 2t\Big|_0^{\pi} = 0\end{aligned}$$

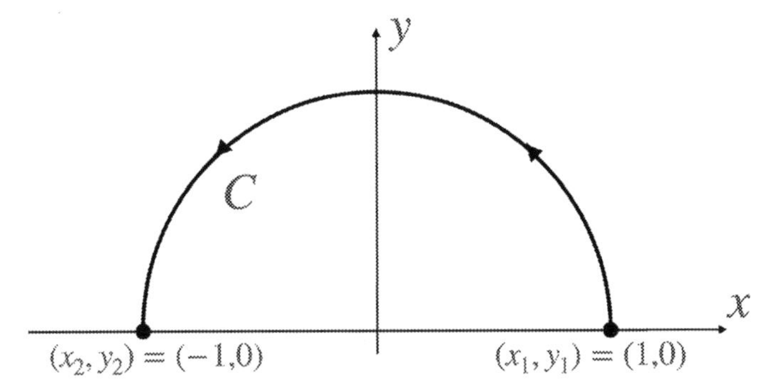

Figure 8.2 A semicircular contour in the complex plane, traced from $(x_1, y_1) = (1,0)$ to $(x_2, y_2) = (-1,0)$.

This might seem like a peculiar result, but there are a number of peculiar results when doing contour integration of complex functions. Some of these peculiar results are very useful for solving problems in Physics.

Now consider the integral along the same contour but for the function

$$f(z) = \frac{1}{z} = \frac{1}{z}\frac{z^*}{z^*} = \frac{x - iy}{x^2 + y^2}$$

The denominator $x^2 + y^2 = 1$ along this contour, and $u = x$ and $v = -y$, so

$$\begin{aligned}
\int_C f(z)\,dz &= \int_0^\pi [\cos t(-\sin t) + \sin t(\cos t)]\,dt \\
&\quad + i\int_0^\pi [-\sin t(-\sin t) + \cos t(\cos t)]\,dt \\
&= 0 + i\int_0^\pi \left[\cos^2 t + \sin^2 t\right]\,dt = i\int_0^\pi dt = \pi i
\end{aligned}$$

and the result is once again peculiar.

Note that the contour we follow makes an important difference. In the second example above, if we went from $(x_1, y_1) = (1,0)$ to $(x_2, y_2) = (-1,0)$ but along a semicircle C traced clockwise through the lower half of the complex plane, we would have found the integral to be $-\pi i$ instead of $+\pi i$.

8.4.1 INTEGRAL AROUND A CLOSED LOOP

It is pretty easy to see that if our contour in the two examples of the previous section was a closed loop, starting and ending at $(x, y) = (1,0)$, along a circular contour C traced counter clockwise, then we would have found

$$\oint_C z\,dz = 0 \qquad \text{and} \qquad \oint_C \frac{1}{z}\,dz = 2\pi i$$

This is actually a hint of two very useful theorems that we will now discuss.

First, rewrite (8.7) in terms of two real line integrals in the (x,y) plane using $\vec{A} \equiv u\hat{i} - v\hat{j}$ and $\vec{B} \equiv v\hat{i} + u\hat{j}$ as

$$\oint_C f(z)\,dz = \oint_C \vec{A}\cdot d\vec{r} + i\oint_C \vec{B}\cdot d\vec{r}$$

If C is a closed contour, then we can write each of these two integrals using Stoke's Theorem (Section 4.3.1). Applying the Cauchy-Riemann relations (8.6), we find

$$\oint_C \vec{A}\cdot d\vec{r} = \left[\frac{\partial u}{\partial y} + \frac{\partial v}{\partial x}\right]\hat{k} = 0 \qquad \text{and} \qquad \oint_C \vec{B}\cdot d\vec{r} = \left[\frac{\partial v}{\partial y} - \frac{\partial u}{\partial x}\right]\hat{k} = 0$$

This proves the *Cauchy-Goursat Theorem*, namely that the integral around any closed contour of any function that is analytic throughout the enclosed region is zero. This explains the result for the example above when $f(z) = z$. We did the integral over a circular contour, but, in fact, it didn't matter whether the contour was circular or not, only that it was closed, since $f(z) = z$ is analytic everywhere.

Note that a corollary of the Cauchy-Goursat Theorem is that the contour integral of an analytic function $f(z)$ between z_1 and z_2 is independent of the contour. This is easy to see. If one of the contours from z_1 to z_2 is reversed, so goes from z_2 to z_1, then the integral picks up a minus sign. We now have a closed contour, that is $z_1 \to z_2 \to z_1$, so the sum of the two integrals must be zero. That is, the two integrals $z_1 \to z_2$ must equal each other.

Of course, the reason we don't get zero when $f(z) = 1/z$ is that this function is not analytic at $z = 0$. In this case, however, there is the *Cauchy Integral Theorem* which states that

$$\oint_C \frac{f(z)}{z - z_0}\,dz = 2\pi i f(z_0) \tag{8.9}$$

where C is a closed contour containing the point $z = z_0$ and $f(z)$ is analytic inside C. As usual, we assume the contour is traced in a counter clockwise direction.

It is not hard to prove (8.9). In fact, we've already done most of the work. We can replace any arbitrary contour C' that encloses $z = z_0$ with a small circular contour C_0 about the pole, plus the contour C shown in Figure 8.3. The integrals along $A \to B$ and $B \to A$ cancel each other out, as does the circular part of C with C_0, so we can replace the integral around C' with the integral around C plus the integral around C_0. The integral around C is zero because there is no pole inside it. Since the radius of C_0 is arbitrarily small, we can take $f(z_0)$ out of the integral, leaving us with the integral of $1/(z - z_0)$, which we have already shown to be equal to $2\pi i$, after translating the axis to put z_0 at the origin. If you want to be more formal about it, write $z - z_0 = re^{i\theta}$ and integrate over $0 \leq \theta \leq 2\pi$ with $dz = ire^{i\theta}\,d\theta$. This proves (8.9).

8.4.2 PRACTICAL MATHEMATICAL EXAMPLES

The Cauchy Integral Theorem (8.9) makes it possible to do many definite integrals which might otherwise seem to be intractable. It is best to illustrate this with a specific example, although different examples might use different, but similar, approaches.

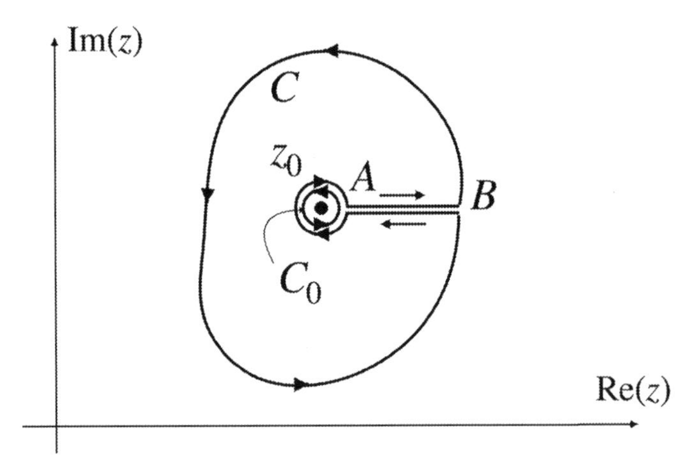

Figure 8.3 Two closed contours C and C_0 in the complex plane. We argue that any contour C', where the opening at point B is closed and there is no "leg" over to point A, can be replaced by the contours C, which does not include the pole at z_0, plus C_0, which is circular and has an arbitrarily small radius and can therefore be easily evaluated.

Consider the integral

$$\mathscr{I} = \int_0^\infty \frac{\cos x}{1+x^2}\,dx \tag{8.10}$$

which in fact can be written as a contour integral of the form (8.9) in which case it is simple to read off the answer. First, realize that the integrand is an even function, and that if I replace the cosine with a sine function, it would be an odd function. This means that

$$\mathscr{I} = \frac{1}{2}\int_{-\infty}^{\infty} \frac{\cos x}{1+x^2}\,dx = \frac{1}{2}\int_{-\infty}^{\infty} \frac{\cos x}{1+x^2}\,dx + i\frac{1}{2}\int_{-\infty}^{\infty} \frac{\sin x}{1+x^2}\,dx = \frac{1}{2}\int_{-\infty}^{\infty} \frac{e^{ix}}{1+x^2}\,dx$$

The final form can be equated to a contour integral, using the contour shown in Figure 8.4. The contour C runs along the x-axis in the positive direction, for some finite length symmetric with the y-axis, and closes with a semicircle in the *upper* hemisphere, returning to the x-axis. We choose the upper hemisphere because we want the integrand to go to zero as the radius of the semicircle goes to infinity. The denominator of the integrand tends to infinity for either the upper or lower hemisphere, but the numerator only goes to zero in the upper hemisphere. This is because the imaginary part of z is large and positive, so the exponential e^{ix} becomes small.

Furthermore, as the radius of C goes to infinity, the contour includes the entire real axis. In other words, for the contour C in Figure 8.4,

$$\mathscr{I} = \frac{1}{2}\oint_C \frac{e^{iz}}{1+z^2}\,dz$$

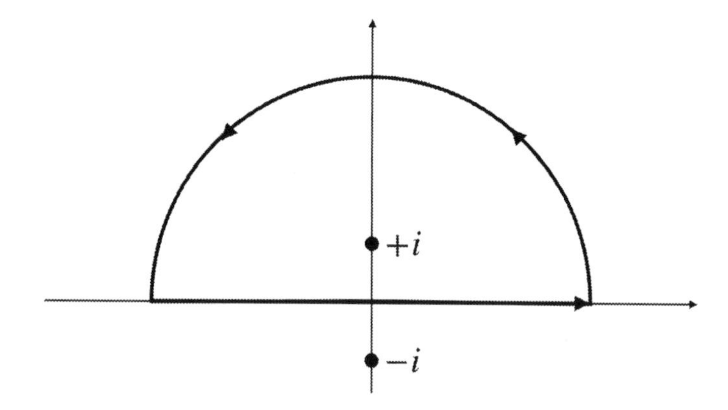

Figure 8.4 A contour in the complex plane used to evaluate the integral (8.10). The contour is traced counter clockwise, and the radius of the semicircle is taken to infinity. The integrand goes to zero along the semicircle, leaving only the contribution from the x-axis. However, the integral is actually evaluated using (8.9) and the pole at $z_0 = +i$.

which is straightforward to evaluate using (8.9). We have

$$\mathscr{I} = \frac{1}{2}\oint_C \frac{e^{iz}}{(z+i)(z-i)}\,dz = 2\pi i \frac{1}{2}\left.\frac{e^{iz}}{z+i}\right|_{z=+i} = \frac{\pi}{2e}$$

EXERCISES

8.1 *If $z = x+iy$, where x and y are real numbers, prove that the function $f(z) = e^z$ is analytic everywhere in the complex plane.*

8.2 *If $z = x+iy$, where x and y are real numbers, prove that the function $f(z) = 1/z$ is analytic everywhere in the complex plane except at $z = 0$.*

8.3 *This exercise involves functions that are not analytic everywhere.*

(a) *A complex function $f(z) = 2y + ix$ where $z = x+iy$. Use the definition of the derivative directly to show that $f'(z)$ does not exist anywhere in the complex plane. Then show that this is consistent with the Cauchy-Riemann relations.*
(b) *A complex function $f(z) = |x| - i|y|$ where $z = x+iy$. Where, if anywhere, in the complex plane is this function analytic?*

8.4 *This exercise involves functions that are analytic.*

(a) *Prove that $f(z) = e^z$ is analytic everywhere in the complex plane.*
(b) *Show that if $f(z)$ is an analytic function of z, then $g(z) = zf(z)$ is also an analytic function of z. Use this to explain why*

$$f(z) = \sum_{n=0}^{\infty} c_n z^n$$

is an analytic function of z.

8.5 *Find an analytic function $f(z) = u(x,y) + iv(x,y)$ whose imaginary part is*

$$v(x,y) = (y\cos y + x\sin y)e^x$$

8.6 *For each of the two paths (a) and (b) in this figure, where (b) is a horizontal step followed by a vertical step,*

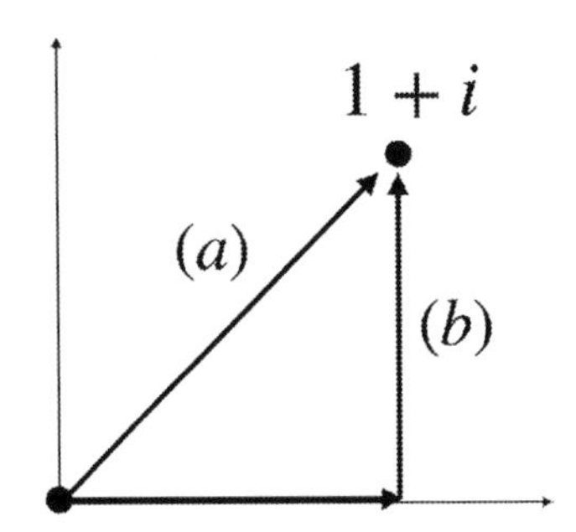

calculate the integral

$$\int_0^{1+i} (z^2 - z)\, dz$$

Explain why the two results compare to each other the way that they do.

8.7 *Calculate the integral*

$$\oint_C \frac{z^2}{2z-1}\, dz$$

for the following closed contour C, a square of side length 2 centered on the origin:

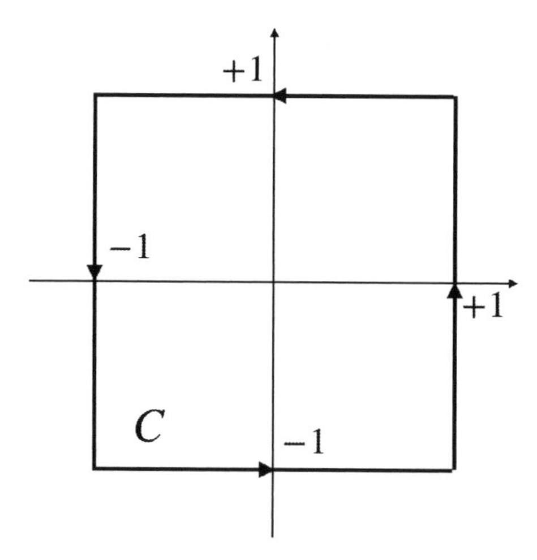

8.8 *By direct integration, calculate the integral*

$$\mathscr{I} = \int_C \frac{1}{z}\, dz$$

around the square contour with side length 2a shown here:

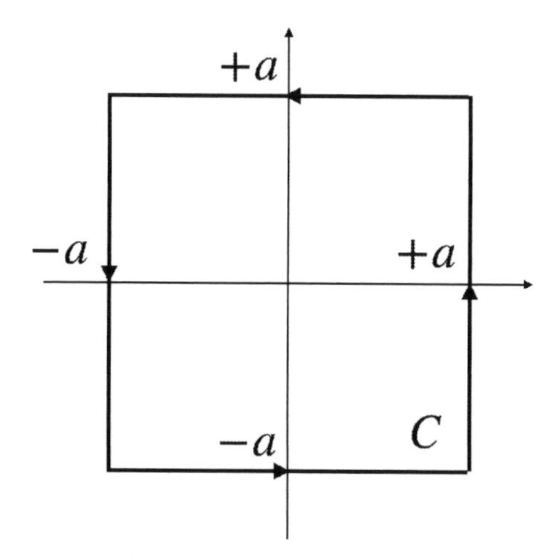

and compare to the result you get from the Cauchy Integral Theorem.

8.9 *Evaluate the integral*

$$\mathscr{I} = \int_{-\infty}^{\infty} \frac{e^{ikx}}{4x^2+1}\, dx$$

separately for the cases $k > 0$ and $k < 0$.

8.10 *Calculate the integral*

$$\int_0^{\infty} \frac{1}{x^2+1} dx$$

in two different ways. First, use the substitution $x = \tan\theta$*, and second as a contour integral in the complex plane.*

8.11 *Use contour integration to find the Fourier transform* $A(k)$ *of the function*

$$f(x) = \frac{C}{x^2 + a^2}$$

where C *and* a *are positive constants.*

8.12 *This exercise is an introduction to an important application of complex functions, namely "conformal maps." Consider two complex variables* $w = u + iv$ *and* $z = x + iy$*, and the "map" given by* $w = f(z)$*. If I draw lines of constant* u *and constant* v*, they are clearly perpendicular to each other in the* (u, v) *plane.*

(a) *For the map* $w = z^2$*, draw the contours* $u = \text{constant}$ *and* $v = \text{constant}$ *in the* (x, y) *plane. Overlap the two sets of contours on the same plot. It should be apparent that wherever they cross, they, again, are perpendicular to each other.*
(b) *By finding the slopes* dy/dx *at any point* (x, y)*, show that the contours are indeed perpendicular to each other.*
(c) *Prove that these contours are perpendicular to each other for a map based on any analytic function.*

Conformal maps find uses in many areas of science and engineering.

9 Probability and Statistics

The notions of probability and probability distributions are central to many fields of physics. Most notable of these are statistical thermodynamics and quantum mechanics, not to mention experimental physics in general. This chapter will lay down some of the basic mathematics associated with these areas.

You might also consider this chapter to be an introduction to the mathematics of "randomness" in physics. This includes constructing the probability of a certain outcome out of a set of lower level outcomes that are equally probability (like throwing a certain roll of dice), as well as making use of artificial randomness, as you would generate on a computer, to calculate mathematical and physical properties.

If there are a number n of potential different outcomes from some experiment or measurement, and all of these different outcomes are equally probable, then coming up with any one of them has a probability of just $1/n$. If the outcomes are not equally probable, then we speak of a *probability distribution* $\mathscr{P}$ of any one of those outcomes. This distribution is some function of the different outcomes, labeled in whichever way is appropriate. If I sum this function over the different possible outcomes (or integrate it if the outcomes are continuous) then you have to get unity. For example, for n equally probable outcomes i, $\mathscr{P}(i) = 1/n$.

When you do an experiment, in which the outcomes of some measurement can be sorted numerically, a natural way to present the data is using a *histogram*. Unless the amount of data is very large, the histogram will be an approximation to the *shape* of the distribution. It is an approximation because there will be some statistical fluctuations in each of the groups of measurements.

We are going to start this chapter by "throwing dice." This is a good way to illustrate all of these concepts. After that, we'll get more formal with the mathematics, and then extend these concepts.

9.1 THROWING DICE

Dice are a prototype for learning about probability and probability distributions. Dice are little cubes – the singular is "die" – with each side labeled with one through six dots. See Figure 9.1. If you throw any one die (and it is "fair") then there is an equal probability of getting any of the faces to land up. That is, the probability of getting a "one" is $1/6$, the same as the probability of getting a "five," and so on.

Now imagine that you throw a handful of six dice at the same time, count the number of dice that land with a "one" face up, and then repeat the experiment several or many more times. The result of each throw will be a number between zero and six. It seems likely that one or two of the dice might land with a "one" face up, but it seems very unlikely that, for example, all six will land this way. Our goal is to try to understand these probabilities mathematically.

DOI: 10.1201/9781003355656-9

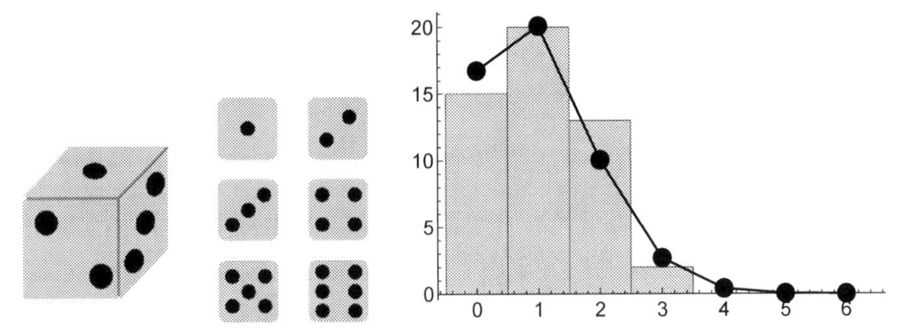

Figure 9.1 Dice are cubes with each of six faces labeled with the number of dots. The right shows the result of an experiment (histogram), compared to the expected, that is, average, result (dots), of throwing a handful of six dice and counting the number of them that land with a "one" face up.

Figure 9.1 also shows the result of such an experiment, throwing fifty handfuls of six dice and each time counting the number that land with a "one" facing up. (Actually, I did not throw fifty handfuls of dice, but instead generated the data using a computer with a random number facility.) The result of the experiment is plotted as a *histogram*, where a bar is drawn at each "bin" showing the "frequency" of the result corresponding to that bin. In other words, 20 of the 50 throws had exactly one die landing with a "one" facing up, 13 had two dice, and so on.

Indeed, our expectations seem to be borne out. It often happened that one or two, or zero, dice landed that way. However, it was much rarer that three dice landed with a "one" face up, and it never happened that more than three came up. Presumably, if I threw ten or a hundred times as many handfuls, then I would have gotten a few to land with ones on five or six dice.

Let's try to predict the outcome that exactly one of the six dice lands with a "one" facing up. Let $p = 1/6$ be the probability of any one die landing with a "one" facing up. Then $q = 1 - p = 5/6$ is the probability the other five dice landing otherwise. Your first instinct might be to say that the probability of exactly one of the dice to land face up is $(1/6)^1(5/6)^5$. However, there are six equivalent ways you can get that result, since all six dice are identical. Therefore, we expect the number of throws that give us exactly one dice with a "one" facing up is the total number of throws times the probability, namely

$$N(1) = N_{\text{total}} \times \mathscr{P}(1) = 50 \times 6 \left(\frac{1}{6}\right)^1 \left(\frac{5}{6}\right)^5 = \frac{78125}{3888} \approx 20.09$$

which is pretty close to the number (20) that were observed. Of course, you can never observe exactly 20.09 occurrences because the answer has to be an integer, but, on average, this is presumably what you'd find.

Now let's predict the number of throws where exactly two dice land "successfully" with "ones" facing up. The probability of this occurrence for specific dice is now $(1/6)^2(5/6)^4$, but counting up the number of combinations that can land this way is a little trickier. We need to ask ourselves, "How many ways are there of picking two dice out of a set of six?"

Well, I have six ways to pick the first of the two, and then five ways to pick the second of the two, but I have to be aware of "double counting." That is, I could pick die #3 first and then die #5, but that's the same result as picking die #5 first and then die #3. So, the answer is $6 \cdot 5/2 = 15$. Therefore, we expect the number of successful throws to be

$$N(2) = N_{\text{total}} \times \mathscr{P}(2) = 50 \times 15 \left(\frac{1}{6}\right)^2 \left(\frac{5}{6}\right)^4 = \frac{78125}{7776} \approx 10.05$$

which again is pretty close to the number (13) that we observed.

Repeating this process for the other bins gives the black points in Figure 9.1, which I've simply joined with a straight line. It seems like we now have an understanding of how to handle probabilities of different outcomes, so we'll formalize the mathematics and then make some extensions.

9.2 COUNTING PERMUTATIONS AND COMBINATIONS

It sounds to be clear from our example with the dice, that we need to learn how to count efficiently. For example, we needed to know the number of ways we could select two dice from a collection of six of them, which turned out to be $6 \cdot 5/2 = 15$. We will now set up the mathematics for these kinds of calculations in general.

If you have n things arranged in a particular order, then the number of possible orderings, called *permutations*, is n factorial, that is

$$n! = n(n-1)(n-2)\cdots 1$$

Try it by arranging the numbers 1, 2, and 3 in all possible orders. You have $1,2,3$ and $3,1,2$ and $2,3,1$ (called "cyclic permutations") and also the three orderings you get by flipping the first two of each of these, namely $2,1,3$ and $1,3,2$ and $3,2,1$. That's a total of $3! = 3 \cdot 2 \cdot 1 = 6$ permutations.

Now suppose you don't care about the order, but you are interested in different subsets of n things. (For example, "How many ways are there of picking two dice out of a set of six?") You now ask what is the number of *combinations* of n things taken m at a time. Well, there are n ways to pick the first thing, then $n-1$ ways to pick the second, and so on, you might write the number of combinations as

$$n(n-1)(n-2)\cdots(n-m+1)$$

which is a product of m factors. However, since I don't care about the order, I could pick the second first, and the third second, and the first third, and..., so I need to divide by the number of ways that I could order m possibilities, namely $m!$. Therefore, the number of combinations is

$$\binom{n}{m} \equiv \frac{n(n-1)(n-2)\cdots(n-m+1)}{m!} = \frac{n!}{m!(n-m)!} \tag{9.1}$$

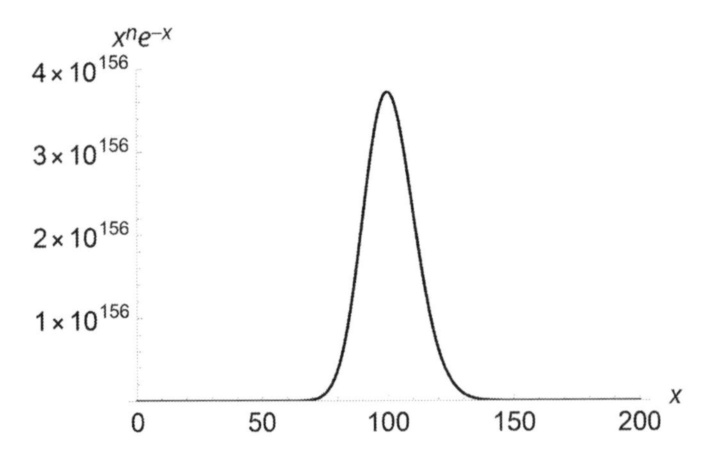

Figure 9.2 The integrand $x^n e^{-x}$ of (9.2) for $n = 100$, plotted as a function of x. Notice that the integrand peaks at $x = n = 100$, and that the shape resembles that of a Gaussian. Therefore, an approximation to $n!$ can be obtained by integrating over a Gaussian function with the appropriate mean and width.

This symbol is called the *Binomial Coefficient* because it tells you the coefficient of the term $p^m q^{n-m}$ in the expansion of $(p+q)^n$, namely the number of ways you can get m factors of q when you expand $(p+q)^n$. (See Section 9.3.1.)

Note that for throwing six dice, the number of combinations for one and two successes are

$$\binom{6}{1} = \frac{6!}{1!5!} = 6 \qquad \text{and} \qquad \binom{6}{2} = \frac{6!}{2!4!} = 15$$

which is just what we got in the previous section.

9.2.1 LARGE NUMBERS AND STIRLING'S APPROXIMATION

Equation (9.1) is key to determining many quantities related to probabilities, and this includes almost the entire field of Statistical Mechanics.[1] However, the number of "dice" in those examples is very large, on the order of Avogadro's Number, namely $N_0 = 6.02 \times 10^{23}$. The form (9.1) is not handy for dealing with numbers like this, and we would really prefer to have some closed analytic form for $n!$.

Fortunately, we have the Γ-function (1.13) from Section 1.5.5, which we reproduce here as

$$\Gamma(n+1) = \int_0^\infty x^n e^{-x}\, dx = n! \tag{9.2}$$

for a non-negative integer $n = 0, 1, 2, 3, \ldots$. Figure 9.2 plots the integrand as a function of x for $n = 100$. The peak position and a shape that resembles a Gaussian

[1] An excellent reference is *An Introduction to Thermal Physics*, by Daniel V. Schroeder, Oxford University Press; 1st edition (2021). This derivation follows Appendix B.3.

suggest that we can use what we know about Gaussian integrals to come up with an approximation for (9.2) when n is large.

First let's establish the position of the peak in Figure 9.2. Since

$$\frac{d}{dx}x^n e^{-x} = nx^{n-1}e^{-x} - x^n e^{-x} = x^{n-1}e^{-x}(n-x) = 0$$

it is apparent that the Gaussian approximation peaks at $x = n$. We then define $y = x - n$ and write

$$\begin{aligned} x^n e^{-x} &= e^{n\log x}e^{-x} = e^{n\log(y+n)}e^{-(y+n)} \\ &= e^{n\log[n(1+y/n)]}e^{-n}e^{-y} = e^{n\log n}e^{n\log(1+y/n)}e^{-n}e^{-y} \end{aligned}$$

Since n is a large number, and it is clear from Figure 9.2 that the integrand is very small if $|x-n|$ is large, we can Taylor expand $\log(1+y/n)$ using (2.8) to get

$$x^n e^{-x} \approx e^{n\log n}e^{-n}e^{n(y/n - y^2/2n^2)}e^{-y} = e^{n\log n}e^{-n}e^{-y^2/2n}$$

and we have confirmed that the integrand is a Gaussian in y, centered at $y = 0$. Finally, since the width of this Gaussian is on the order of $\sqrt{n} \gg 1$, we can replace the lower limit of the integral with $-\infty$, and use (1.14) to evaluate it. This results in

$$n! \approx e^{n\log n}e^{-n}\int_{-\infty}^{\infty} e^{-y^2/2n}\,dy = e^{n\log n}e^{-n}\sqrt{2\pi n} = n^n e^{-n}\sqrt{2\pi n} \qquad (9.3)$$

which is known as *Stirling's Approximation*. An alternate form using the logarithm is

$$\log n! \approx n\log n - n + \frac{1}{2}\log(2\pi n)$$

where the third term is typically neglected for $n \sim 10^{23}$ in statistical mechanics.

Figure 9.3 shows the agreement between (9.3) and the exact value for $n!$, for n up to 100. The approximation is reasonably good, but gets much better for numbers n that are very large, on the order of Avogadro's Number. This is the reason it is very useful in Statistical Mechanics.

We will also find Stirling's Approximation useful when we discuss the Gaussian Probability Distribution Function, Section 9.3.3.

9.3 PROBABILITY DISTRIBUTIONS

If a function $\mathscr{P}(m)$ gives the probability of measuring outcome m, where m is an integer, then we call $\mathscr{P}(m)$ a *discrete probability distribution function*. If the outcome is described by a continuous variable x, then $\mathscr{P}(x)\,dx$ gives the probability of measuring an outcome between x and $x + dx$, and $\mathscr{P}(x)$ is a *continuous probability distribution function*. Generically, we call $\mathscr{P}$ a *distribution*. Probability distributions are important throughout physics, from data analysis, to quantum mechanics, to statistical mechanics.

Our approach in this section will be to first write down the *binomial distribution*, which describes the situation with throwing dice. We will then discuss two extensions of the binomial distribution, namely the *Poisson distribution* and the *Gaussian distribution*, also known as the *normal distribution*.

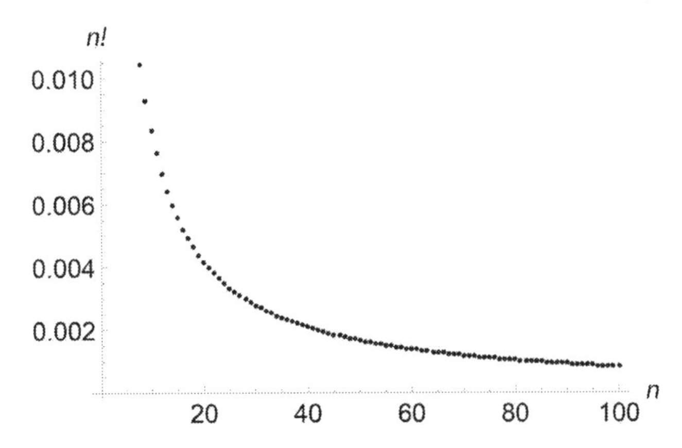

Figure 9.3 Plot of the fractional difference between $n!$ and Stirling's Approximation (9.3) for n up to 100. The approximation is rather good at $n = 100$, differing from the exact value only by about 0.1%, but the real value of Stirling's Approximation for physics is for very large n, on the order of Avogadro's Number 6.02×10^{23}.

9.3.1 THE BINOMIAL DISTRIBUTION

We now return to our problem of throwing a handful of dice, and generalize to the question, "What is the probability that if I throw n dice, then m of them land with a one facing up?" The probability $p = 1/6$ that any one of the dice will land this way. Therefore, the probability that throwing a handful of n dice and having m of them land "successfully" with "ones" facing up is

$$\mathscr{P}_{\text{bin}}^{(1/6)}(m) = \binom{n}{m}\left(\frac{1}{6}\right)^m\left(\frac{5}{6}\right)^{n-m}$$

It is obvious how to generalize this to any situation where the probability of an individual success is p, namely

$$\mathscr{P}_{\text{bin}}^{(p)}(m) = \binom{n}{m} p^m q^{n-m} \tag{9.4}$$

where $p+q=1$. As mentioned in Section 9.2, this is called the *binomial distribution* because it gives the term proportional to p^m of the expansion $(p+q)^n$. As such, it is clear, then, that

$$\sum_{m=0}^{n} \mathscr{P}_{\text{bin}}^{(p)}(m) = (p+q)^n = 1^n = 1$$

which must be the case, of course, since the only possible results for m are the integers from zero to n. We say that the probability distribution is *normalized*.

The black points on the right in Figure 9.1 are $N\mathscr{P}_{\text{bin}}^{(1/6)}(m)$ where m is labeled along the horizontal axis. Given a number N of dice throws, the number you expect to get would be $N\mathscr{P}_{\rm bin}^{(1/6)}(m)$ for m dice landing with a "one" facing up. The histogram is an approximation to $\mathscr{P}_{\rm bin}^{(1/6)}(m)$, after correcting for the normalization.

9.3.2 THE POISSON DISTRIBUTION

The *Poisson distribution* is a limiting case of the binomial distribution, and is often taken for granted as *the* correct distribution for many problems in the statistical analysis of data. The reason is that this limiting case, namely $n \to \infty$ and $p \to 0$ with $\mu = np$ kept finite, is so often the case in physical systems. The classic case is radioactive decay, where the probability of any one radioactive nucleus decaying in a certain time interval is very small, but the number of possible decaying nuclei is very large. Another example might be the number of people in Philadelphia who might call Temple University on a Tuesday afternoon between 3pm and 4pm.

If we write (9.4) as

$$\mathscr{P}_{\rm bin}^{(p)}(m) = \frac{n!}{m!(n-m)!}p^m(1-p)^{n-m}$$

then we can see how to take the limit $n \to \infty$ and $p \to 0$ with $\mu = np$ kept finite. Firstly

$$\lim_{n\to\infty}\frac{n!}{(n-m)!} = \lim_{n\to\infty} n(n-1)\cdots(n-m+1) = n^m$$

which gives a factor $n^m p^m = \mu^m$. Now we also have

$$\lim_{n\to\infty}\lim_{p\to 0}(1-p)^{n-m} = \lim_{n\to\infty}\lim_{p\to 0}(1-p)^n = \lim_{p\to 0}(1-p)^{\mu/p} = \lim_{p\to 0}(1-p)^{-\mu/(-p)} = e^{-\mu}$$

(You might want to recall how we came up with e in Section 1.5.2.) Putting this together gives us the Poisson distribution, namely

$$\mathscr{P}_{\rm Poiss}^{(\mu)}(m) = \frac{\mu^m}{m!}e^{-\mu} \tag{9.5}$$

It is easy to confirm that the Poisson distribution is properly normalized, just by summing

$$\sum_{m=0}^{n}\mathscr{P}_{\rm Poiss}^{(\mu)}(m) \to \sum_{m=0}^{\infty}\mathscr{P}_{\rm Poiss}^{(\mu)}(m) = \left[\sum_{m=0}^{\infty}\frac{\mu^m}{m!}\right]e^{-\mu} = e^{\mu}e^{-\mu} = 1$$

Figure 9.4 plots (9.5) for three different values of μ. Note that for low values of μ, the probability of $m = 0$ is significant. As μ increases, the distribution peaks close to, and becomes more symmetric about, the mean.

One of the most useful properties of the Poisson distribution is that the standard deviation (Section 9.4.1) $\sigma = \sqrt{\mu}$. When you are doing an experiment and you expect the number of "successes" to be Poisson distributed, the number you observe is your best estimate of what the average out to be, so you quote a "standard error" that is the square root of the number you observe.

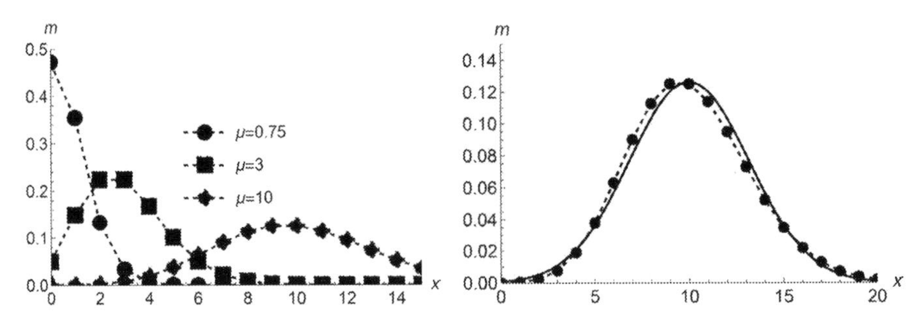

Figure 9.4 Three examples of a Poisson distribution, with three different values for the mean $\mu = 0.75, 3, 10$, are plotted on the left. On the right we reproduce the Poisson distribution for $\mu = 10$, along with the Gaussian distribution (solid line) for the same μ and $\sigma = \sqrt{\mu}$.

9.3.3 THE GAUSSIAN DISTRIBUTION

The Gaussian (or normal) probability distribution function is probably the most common example of a continuous variable x. It is widely used in data analysis and other fields of physics, and can be derived from the binomial or Poisson distributions in the case where $m \gg 1$ and you don't stray too far from the average value. Below we will see how do this with the Poisson distribution. The result is

$$\mathscr{P}^{(\mu,\sigma)}_{\text{Gauss}}(x) = \frac{1}{\sigma\sqrt{2\pi}} e^{-(x-\mu)^2/2\sigma^2} \tag{9.6}$$

It should be clear from Section 1.5.6 that this form is normalized so that

$$\int_{-\infty}^{\infty} \mathscr{P}^{(\mu,\sigma)}_{\text{Gauss}}(x)\, dx = 1$$

Let's see now how the Gaussian distribution arises from the Poisson distribution for $m \gg 1$ and $\mu \gg 1$. Take the logarithm of (9.5) and use Stirling's approximation (9.3) to write

$$\begin{aligned} \log \mathscr{P}^{(\mu)}_{\text{Poiss}}(m) &= m \log \mu - \mu - \log m! \\ &= m \log \mu - \mu - m \log m + m - \frac{1}{2}\log(2\pi) - \frac{1}{2}\log m \end{aligned}$$

We know that (9.5) has a peak when $m \approx \mu$, so it makes sense to define a variable $y \equiv m - \mu$ and expand the log around $y = 0$ using (2.8). We have

$$\begin{aligned} \log \mathscr{P}^{(\mu)}_{\text{Poiss}}(m) &= (\mu + y)\log \mu + y - \left(\mu + y + \frac{1}{2}\right)\log(\mu + y) - \frac{1}{2}\log(2\pi) \\ \text{with} \qquad \log m &= \log(y + \mu) = \log \mu + \log\left(1 + \frac{y}{\mu}\right) \approx \log \mu + \frac{y}{\mu} - \frac{y^2}{2\mu^2} \end{aligned}$$

Some algebra that is more tedious than tricky gives

$$\begin{aligned}\log \mathscr{P}_{\text{Poiss}}^{(\mu)}(m) &= y - \frac{1}{2}\log\mu - (\mu+y)\frac{y}{\mu} + (\mu+y)\frac{y^2}{2\mu^2} - \frac{y}{2\mu} - \frac{y^2}{2\mu^2} - \frac{1}{2}\log(2\pi)\\ &= \log\frac{1}{\sqrt{2\pi\mu}} - \frac{y^2}{\mu} + \frac{y^2}{2\mu} + \frac{y^3}{2\mu^2} - \frac{y}{2\mu} - \frac{y^2}{2\mu^2}\\ &= \log\frac{1}{\sqrt{2\pi\mu}} - \frac{y^2}{2\mu} + \frac{y^3}{2\mu^2} - \frac{y}{2\mu} - \frac{y^2}{2\mu^2}\end{aligned}$$

Rewriting this in terms of $\sigma = \sqrt{\mu}$ gives us

$$\log \mathscr{P}_{\text{Poiss}}^{(\mu)}(m) = \log\frac{1}{\sigma\sqrt{2\pi}} - \frac{y^2}{2\sigma^2} + \frac{1}{\sigma}\frac{y^3}{2\sigma^3} - \frac{1}{\sigma}\frac{y}{2\sigma} - \frac{1}{\sigma^2}\frac{y^2}{2\sigma^2}$$

If $y/\sigma = y/\sqrt{\mu}$ is not very large, that is if we stay close to the mean, and $\sigma = \sqrt{\mu} \gg 1$, we can neglect the last three terms compared to the second. The result is

$$\log \mathscr{P}_{\text{Poiss}}^{(\mu)}(m) \to \log\frac{1}{\sigma\sqrt{2\pi}} - \frac{y^2}{2\sigma^2} \qquad \text{so} \qquad \mathscr{P}_{\text{Poiss}}^{(\mu)}(m) \to \frac{1}{\sigma\sqrt{2\pi}}e^{-(m-\mu)^2/2\sigma^2}$$

which is the Gaussian approximation. Figure 9.4 also compares the Poisson distribution for $\mu = 10$ with the Gaussian distribution for the same mean and $\sigma = \sqrt{\mu}$. The agreement is obviously very good, even though $\sqrt{10}$ is not so much larger than unity. You might notice, however, that the fractional disagreement on the tails of the distribution, that is when we are far from the mean, are rather large.

9.4 BASIC DATA ANALYSIS

The term "data" is impossibly broad, so we're going to have a discussion here of simple topics confined to a simple definition of "data." It will illustrate some of the most important concepts, though, that you are likely to be concerned with during the studies of physics.

By "data" we mean generic sets of real numbers. For example, we can label them x_i and y_i for $i = 1,2,3,\ldots,N$. We won't make use of it here, but it is commonplace to refer to these data sets as $\underline{x}$ and $\underline{y}$ in the sense of generalized vectors, as in Section 6.2.

The arrays $\{x_i\}$ and $\{y_i\}$ are individually sets of data, and we will talk about how to calculate important quantities that collective describe them. It is also reasonable to think of these as correlated somehow, in which case we think of the data as $\{x_i, y_i\}$. In this case, it is common to have some model $y = f(x)$ which describes how the two data sets depend on each other. In many applications, the function $f(x)$ will have so-called "free parameters" which can be fit to describe the data, and there are well established techniques for finding the best fit.

9.4.1 MEAN, VARIANCE, AND STANDARD DEVIATION

A very familiar concept that you would use to describe a set of data $\{x_i\}$ is the *mean* or *average* of the values. Notations for the mean include $\langle x\rangle$ (which I will use), $\bar{x}$, and x_{avg}. You are likely well aware of the calculation of the mean of N values $\{x_i\}$, namely

$$\bar{x} = \langle x \rangle = \frac{1}{N}\sum_{i=1}^{N} x_i \tag{9.7}$$

A less familiar concept, but at least as important as the mean, is the *variance* σ^2, which describes the "spread" of the values around the mean. It is simply the average of the square of the deviations of the values from the mean. In other words

$$\sigma^2 = \langle (x - \langle x \rangle)^2 \rangle = \langle x^2 \rangle - 2\langle x \rangle\langle x \rangle + \langle x \rangle^2 = \langle x^2 \rangle - \langle x \rangle^2 \tag{9.8}$$

The square root of the variance, that is σ, is called the *standard deviation.*

When the values $\{x_i\}$ are distributed according to some established probability distribution function, then it is possible to quantitatively predict the probability that the mean $\langle x \rangle$ is within some number of standard deviations from the correct value. A measurement of x is often reported as $\langle x \rangle \pm \sigma$, but the interpretation of the $\pm$ clearly depends on what is the appropriate distribution function. More often than not, however, people will assume it is a Gaussian distribution where $\langle x \rangle$ is a good approximation to the mean μ of the distribution.

9.4.1.1 Application to binomial, Poisson, and Gaussian distributions

If the values $\{x_i\}$ are nonnegative integers $\{m_i\}$, then it is possible that they are distributed according to the binomial or Poisson distributions. If the $\{x_i\}$ are any real numbers, however, it might be natural to assume that are distributed according to a Gaussian. In these cases, we can calculate directly what we expect for the mean and variance. Of course, there are many other potential distribution functions, but we are only going to consider these three possibilities.

The mean value of some discrete measurement m that is distributed according to the function $\mathscr{P}(m)$ is simply the sum of the values of the measurement times the probability for getting that measurement, namely

$$\mu = \langle m \rangle = \sum_m m\mathscr{P}(m)$$

and the variance is

$$\sigma^2 = \langle (m - \mu)^2 \rangle = \langle m^2 \rangle - \langle m \rangle^2$$

We argued above that the mean of a set of numbers that follow the binomial distribution was np, where n is the number of chances of success, and p was the probability of a single success. This makes sense, because if you have n chances with each a probability p, you expect the average to be np. Let's prove this statement.

In the case the binomial distribution, we have

$$\begin{aligned}\langle m\rangle &= \sum_{m=0}^{n} m\mathscr{P}_{\text{bin}}^{(p)}(m) = \sum_{m=0}^{n} m\frac{n!}{m!(n-m)!}p^m q^{n-m} = \sum_{m=1}^{n} m\frac{n!}{m!(n-m)!}p^m q^{n-m} \\ &= \sum_{k=0}^{n-1} \frac{n!}{k!(n-k-1)!}p^{k+1}q^{n-k-1} = np\sum_{k=0}^{n-1}\frac{(n-1)!}{k!(n-1-k)!}p^k q^{n-1-k} \\ &= np\sum_{k=0}^{l}\frac{l!}{k!(l-k)!}p^k q^{l-k} = np(p+q)^l = np \qquad (9.9)\end{aligned}$$

where $l = n-1$ and $q = 1-p$. The "obvious" result is in fact correct. In order to calculate the variance, we need to evaluate

$$\begin{aligned}\langle m^2\rangle &= \sum_{m=0}^{n} m^2\frac{n!}{m!(n-m)!}p^m q^{n-m} = np\sum_{k=0}^{l}(k+1)\frac{l!}{k!(l-k)!}p^k q^{l-k} \\ &= np\{lp+1\} = np\{(n-1)p+1\} = n^2p^2 - np^2 + np\end{aligned}$$

Therefore, the variance is

$$\sigma^2 = n^2p^2 - np^2 + np - (np)^2 = np(1-p) \qquad (9.10)$$

for the binomial distribution (9.4).

We built the Poisson distribution based on the mean $\mu = np$ of the binomial distribution, where $n \to \infty$ and $p \to 0$. It is therefore clear that

$$\sigma^2 = \mu \qquad (9.11)$$

for the Poisson distribution (9.5).

For a continuous distribution $\mathscr{P}(x)$, we integrate to get the mean and variance, that is

$$\langle x\rangle = \int_{-\infty}^{\infty} x\mathscr{P}(x)\,dx \qquad \text{and} \qquad \langle x^2\rangle = \int_{-\infty}^{\infty} x^2\mathscr{P}(x)\,dx$$

It is easy to apply these to the Gaussian distribution (9.6). Using $y = x-\mu$ we get

$$\begin{aligned}\langle x\rangle &= \int_{-\infty}^{\infty} x\mathscr{P}_{\text{Gauss}}^{(\mu,\sigma)}(x)\,dx \\ &= \frac{1}{\sigma\sqrt{2\pi}}\int_{-\infty}^{\infty} xe^{-(x-\mu)^2/2\sigma^2}\,dx = \frac{1}{\sigma\sqrt{2\pi}}\int_{-\infty}^{\infty}(\mu+y)e^{-y^2/2\sigma^2}\,dy \\ &= \mu\left[\frac{1}{\sigma\sqrt{2\pi}}\int_{-\infty}^{\infty} e^{-y^2/2\sigma^2}\,dy\right] + \frac{1}{\sigma\sqrt{2\pi}}\int_{-\infty}^{\infty} ye^{-y^2/2\sigma^2}\,dy = \mu \qquad (9.12)\end{aligned}$$

where the integral in square brackets is unity because it is just the integral over the probability distribution, and the second integral is zero because the integrand is an odd function of y. Of course, this result is no surprise since we constructed the Gaussian distribution to have a peak at the mean value.

In order to calculate the variance, we need

$$\frac{1}{\sigma\sqrt{2\pi}}\int_{-\infty}^{\infty} x^2 e^{-(x-\mu)^2/2\sigma^2}\,dx = \frac{1}{\sigma\sqrt{2\pi}}\int_{-\infty}^{\infty}(\mu^2+2\mu y+y^2)e^{-y^2/2\sigma^2}\,dy$$

which we consider as three integrals. The first integral is just μ^2, and the second integral is zero because the integrand is odd. For the third integral, use (1.15) to get

$$\int_{-\infty}^{\infty} y^2 e^{-y^2/2\sigma^2}\,dy = \frac{1}{2}\sqrt{\pi 8\sigma^6} = \sigma^3\sqrt{2\pi}$$

Therefore, the variance of the Gaussian distribution (9.6) is

$$\mu^2 + \frac{1}{\sigma\sqrt{2\pi}}\sigma^3\sqrt{2\pi} - \mu^2 = \sigma^2 \tag{9.13}$$

which just goes to show that we used the appropriate notation when we wrote down (9.6). Of course, we had confidence that this was the right answer when we saw how to get the Gaussian distribution from the Poisson distribution.

9.4.2 HISTOGRAMS AS APPROXIMATIONS TO DISTRIBUTIONS

Given a set of data $\{x_i\}$, it is common to reduce them to a *histogram*, which is a convenient way to display the mean and variance pictorially. In fact, the histogram can be considered to be an approximation to the underlying distribution that determines the probabilities for the result of some measurement yielding a specific value, or range of values, of x.

Indeed, the comparison between an experiment and distribution is what we show in Figure 9.1. In this case, each "bin" of the histogram is an integer between 0 and 6, the seven different possible outcomes of our throwing six dice. A distribution function, in this case the binomial distribution, is plotted on top of the histogram to demonstrate agreement between what we observe and what we expect.

Of course, it is also possible to histogram data whose values are continuous. In this case, a histogram of a quantity x is $\Delta N/\Delta x$ where ΔN is the number of events that fall into a "bin" m such that $x_m \le x \le x_m + \Delta x$. (Technically, all of the bins could have a different width Δx_m, but this approach is rarely taken.) When the data values are in fact integers, as in Figure 9.1, it is convenient to set the bin edges to be half-integer.

Figure 9.5 compares the histogram and distribution by multiplying the distribution by the total number of trials N. If, instead, we divide by the histogram by the total number of trials, the result is an approximation to the true distribution, limited by the randomness of the finite number of trials. In the variable x is continuous, then

$$\mathscr{P}(x) = \lim_{\Delta x\to 0}\frac{1}{N}\frac{\Delta N}{\Delta x} = \frac{1}{N}\frac{dN}{dx} \tag{9.14}$$

It should be clear from (9.14) that the integral of $\mathscr{P}(x)$ over all x is unity.

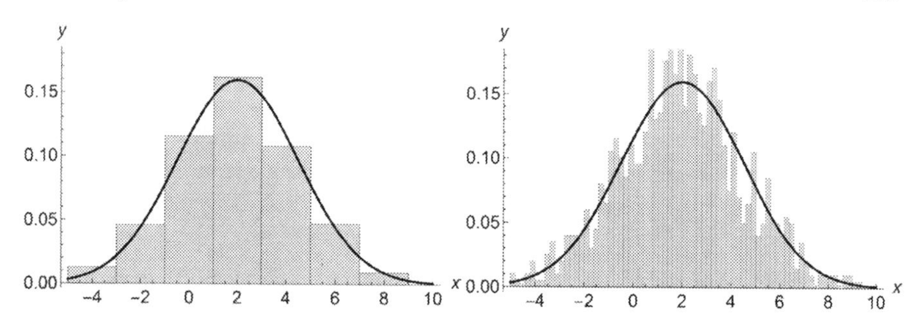

Figure 9.5 Coarse and fine binned histogram examples for data values distributed according to a Gaussian probability distribution with mean 2.0 and standard deviation 2.5. The solid line plots the actual distribution function $\mathscr{P}(x)$.

Figure 9.5 illustrates this concept using a Gaussian probability distribution. A set of 1000 measurements[2] x were taken that should follow a Gaussian distribution with a specific mean and standard deviation. These data are then binned with $\Delta x = 2$ on the left and $\Delta x = 0.2$ on the right, and in each case divided by the number of measurements (i.e. N) and Δx, as indicated in (9.14). In each case, the Gaussian probability distribution is plotted on top of the histogram.

For the coarsely binned histogram, the distribution follows the histogram nearly exactly. This is because each bin has enough entries in it so that its statistical uncertainty is small. The finely binned histogram shows clear statistical variation, with approximately ten times fewer events in each bin that for the coarsely binned version. However, if the object of the experiment is to look for deviations from the predicted distribution, perhaps with a narrow "peak" hiding at some value of x, then the finely binned histogram is the right approach.

9.4.3 FITTING DATA TO MODELS

Given a set of N correlated data points $\{x_i\}$ and $\{y_i\}$, where $i = 1,2,\ldots,N$, is very common to want to "fit" a model function that describes the data as $y = f(x;a_1,a_2,\ldots,a_n)$. That is, we want to find the values of the n parameters a_j, where $j = 1,2,\ldots,n$ so that the functional form passes as close as possible through all the data points. Typically, each data value y_i will have some uncertainty that I'll call σ_i, implying that somehow we know the standard deviation for any individual measurement of y.

If we assume (as everyone pretty much always does) that the points follow a Gaussian distribution, then the probability that data point i comes from the mean determined by the fit function is

$$\mathscr{P}_i(a_1,a_2,\ldots,a_n) = \frac{1}{\sigma_i\sqrt{2\pi}}\exp\left\{-\frac{[y_i - f(x_i;a_1,a_2,\ldots,a_n)]^2}{2\sigma_i^2}\right\}$$

[2]Actually, these values were obtained using a random number generator on a computer. The details of how this is done are discussed in Section 9.5.1.

The probability for the entire set of data is therefore the product of the individual probabilities. This product is called the *likelihood*

$$\mathscr{L}(a_1,a_2,\ldots,a_n)=\prod_{i=1}^{N}\mathscr{P}_i(a_1,a_2,\ldots,a_n)$$

which is messy to write out, so I won't bother. The object of fitting the model to the data then becomes a problem of finding the values of the parameters $a_1,a_2,\ldots,a_n$ which maximize the likelihood.

Multiplying all of those exponentials together will give you a very small number for values of the parameters that are not very close to their final values. This will create headaches with computer applications that try to maximize $\mathscr{L}$, so instead we minimize the negative of the logarithm of the likelihood. That is

$$-\log\mathscr{L}(a_1,a_2,\ldots,a_n)=-\log\prod_{i=1}^{N}\left(\frac{1}{\sigma_i\sqrt{2\pi}}\right)+\frac{1}{2}\chi^2(a_1,a_2,\ldots,a_n)$$

where the first term is just some constant, and the χ^2 function is

$$\chi^2(a_1,a_2,\ldots,a_n)=\sum_{i=1}^{N}\frac{[y_i-f(x_i;a_1,a_2,\ldots,a_n)]^2}{\sigma_i^2} \tag{9.15}$$

The job of finding the best fit parameters has become a problem of finding the set of parameters which minimize $\chi^2(a_1,a_2,\ldots,a_n)$. This procedure is generally referred to as the *method of least squares.*

Given a set of data and some model function $f(x;a_1,a_2,\ldots,a_n)$, a number of tools are available for minimizing $\chi^2(a_1,a_2,\ldots,a_n)$. Most any computer application used for data analysis will have some kind of fitting facility. If you are writing your own code, any respectable numerical library will have other options.

9.4.3.1 Fitting to linear models

The fitting problem reduces to a simple problem in linear algebra if the fitting function is strictly linear in the parameters. That is

$$f(x;a_1,a_2,\ldots,a_n)=a_1g_1(x)+a_2g_2(x)+\cdots+a_ng_n(x)$$

The reason this works out so nicely is that I can take the partial derivatives with respect to the a_j and set them equal to zero to minimize (9.15) and get a system of linear equations to solve for the a_j.

Let's see how this works. The χ^2 function (9.15) becomes

$$\chi^2(a_1,a_2,\ldots,a_n)=\sum_{i=1}^{N}\frac{[y_i-a_1g_1(x_i)-a_2g_2(x_i)-\cdots-a_ng_n(x_i)]^2}{\sigma_i^2}$$

Now take the partial derivative of this with respect to a_k and set it equal to zero.

$$-2\sum_{i=1}^{N}\frac{[y_i-a_1g_1(x_i)-a_2g_2(x_i)-\cdots-a_ng_n(x_i)]g_k(x_i)}{\sigma_i^2}=0$$

This can be rewritten as

$$\sum_{j=1}^{n}\left[\sum_{i=1}^{N}\frac{g_j(x_i)g_k(x_i)}{\sigma_i^2}\right]a_j = \sum_{i=1}^{N}\frac{y_i g_k(x_i)}{\sigma_i^2}$$

which is just a system of linear equations of the form

$$R_{kj}a_j = b_k \qquad \text{where} \qquad R_{kj} = \sum_{i=1}^{N}\frac{g_k(x_i)g_j(x_i)}{\sigma_i^2} \qquad \text{and} \qquad b_k = \sum_{i=1}^{N}\frac{y_i g_k(x_i)}{\sigma_i^2}$$

The coefficients a_j, components of the vector $\underline{a}$, are easily obtained by inverting the matrix $\underline{\underline{R}}$ and multiplying times the vector $\underline{b}$.

Let's work this out for a specific example, namely fitting to the straight line

$$f(x;a_1,a_2) = a_1 + a_2 x$$

In this case, $\underline{\underline{R}}$ is a 2×2 matrix with

$$R_{11} = \sum_{i=1}^{N}\frac{1}{\sigma_i^2} \qquad R_{12} = \sum_{i=1}^{N}\frac{x_i}{\sigma_i^2} = R_{21} \qquad R_{22} = \sum_{i=1}^{N}\frac{x_i^2}{\sigma_i^2}$$

and the best fitting parameters a and b are given by

$$\begin{bmatrix} a_1 \\ a_2 \end{bmatrix} = \begin{bmatrix} R_{11} & R_{12} \\ R_{21} & R_{22} \end{bmatrix}^{-1}\begin{bmatrix} b_1 \\ b_2 \end{bmatrix} = \frac{1}{\Delta}\begin{bmatrix} R_{22} & -R_{12} \\ -R_{21} & R_{11} \end{bmatrix}\begin{bmatrix} b_1 \\ b_2 \end{bmatrix}$$

where $\Delta = R_{11}R_{22} - R_{12}R_{21}$ is the determinant of $\underline{\underline{R}}$ and

$$b_1 = \sum_{i=1}^{N}\frac{y_i}{\sigma_i^2} \qquad \text{and} \qquad b_2 = \sum_{i=1}^{N}\frac{y_i x_i}{\sigma_i^2}$$

Notice that if all of the $\sigma_i = \sigma$, that is, the weights are all equal, then the results for a_1 and a_2 are independent of σ.

9.5 RANDOM NUMBERS AND MONTE CARLO SIMULATIONS

We conclude this section with a very brief discussion of the use of random numbers, generated by some computer application, and their application to carrying out integrals with the Monte Carlo technique.

9.5.1 RANDOM NUMBER GENERATION

Generating random numbers is not as easy as it sounds. Coming up with a computer algorithm that will produce a list of numbers that are truly random is an oxymoron. Nevertheless, a lot of research has been done on how to effectively generate random numbers, and facilities for doing this are built in to just about any computer numerical analysis library or application.

You can take it for granted that computers will generate random numbers that are integers or real numbers, over whatever range you want to specify. Different applications will have different syntax and caveats.

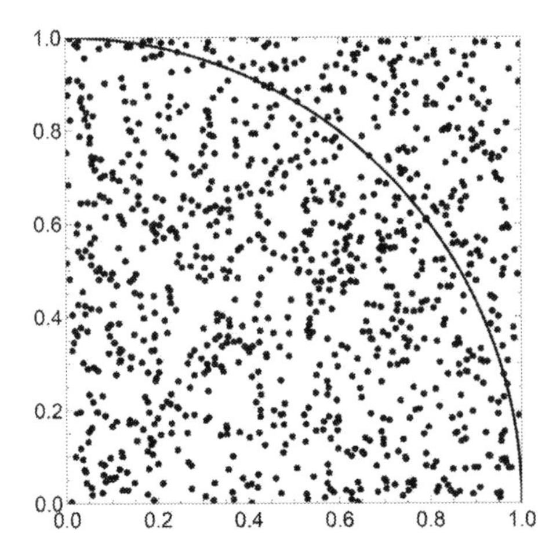

Figure 9.6 Figure showing the Monte Carlo approach to calculating π. The area inside the circular arc is $\pi/4$. This area is estimated by generating random number pairs (x,y) inside the unit square, and then counting the number of pairs for which $x^2+y^2 \leq 1$. An estimate of π comes from multiplying the ratio of those two numbers by four. The figures shows the positions of 1000 generated pairs, 792 of which are inside the arc, giving $\pi \approx 3.168$.

9.5.2 INTEGRATION USING MONTE CARLO TECHNIQUES

You can use random numbers to calculate difficult integrals numerically. These integrals could be over several different variables. The idea is to populate all of the available space defined by the variables, and then count the number of points in this space which satisfy the "area" defined by the integrand.

I will illustrate this with a simple example. The area of one quarter of the unit circle is $\pi/4$, so we can calculate π by integrating under the curve $y = f(x) = \sqrt{1-x^2}$ over the range $0 \leq x \leq 1$. We can estimate the integral by generating a large number N of random points (x,y) over the region of the unit square defined by $0 \leq x \leq 1$ and $0 \leq y \leq 1$, and then counting up the number N_π of points that end up inside the unit circle. As N gets larger and larger, we expect to get better and better approximations to $\pi = 4N_\pi/N$.

This is illustrated in Figure 9.6, which shows the unit circle and 1000 randomly generated points. Counting up all of the points with $x^2+y^2 \leq 1$ gives 792, or $\pi \approx 4 \times 792/1000 = 3.168$. This is very close to $\pi = 3.142$, only 0.84% larger. Generating a large number of events gives a value for π that is closer to the right answer. Running the same code but with 10^5 generated points gives $\pi \approx 3.136$, only 0.18% from the correct value.

EXERCISES

9.1 *Reproduce the histogram in Figure 9.1 and compare your result with the binomial distribution. You can reproduce the histogram by actually throwing a handful* *of six dice a whole bunch of times, and recording the results. You can also use some computer application to generate random numbers and sort them as if they were dice throws. You can also try the same exercise by choosing a number other than six dice to throw each time.*

9.2 *Someone (with large hands) makes 100 throws of handfuls containing eleven eight-sided dice.*

(a) *How many throws should have exactly 3 dice landing with a "seven" facing up?*
(b) *What should be the average number of dice landing with a "seven" facing up?*
(c) *What should be the standard deviation on the average number of dice landing with a "seven" facing up?*

9.3 *The following data comes from a simple pendulum made from a length of string and a weight tied to the end. The table lists the measured the time (in seconds) for* ten swings *for different lengths (in inches) of the string:*

Length	Time
70	26.75
59	24.86
47	21.81
33	18.29
26	16.13
19	13.78
8	8.87

The period T of a pendulum made of a light string of length l and massive bob is $T = 2\pi\sqrt{l/g}$ where g is the acceleration due to gravity. Use this data (or take some data of your own!) and answer the following questions:

(a) *Calculate a value of g for each data point, and compare to the expected value $g = 9.8\ m/s^2$. (You'll of course want to convert the inches to meters.) Calculate the average value $\langle g \rangle$ and the standard deviation σ.*
(b) *Now determine g by doing a straight line fit of the form $y = a_1 + a_2 x$ where $y = T^2$ and $x = l$. In this case, you expect $a_1 = 0$ and $a_2 = 4\pi^2/g$.*

9.4 *Calculate the integral*

$$\int_0^1 x^2\, dx$$

using a Monte Carlo technique. Generate random (x, y) pairs with each distributed between 0 and 1, and count the number of pairs where $y_i \leq x_i^2$. Compare your answer to what you know to be the correct answer.

Index

Printed in the United States
by Baker & Taylor Publisher Services